D1453046

Solid Analytic Geometry

Solid Analytic Geometry

A. Adrian Albert

Eliakim Hastings Moore Distinguished Service
Professor of Mathematics
The University of Chicago

Dover Publications, Inc.
Mineola, New York

Bibliographical Note

This Dover edition, first published in 2016, is an unabridged republication of the work originally published by McGraw-Hill Book Company, Inc., New York, in 1949.

Library of Congress Cataloging-in-Publication Data

Names: Albert, A. Adrian (Abraham Adrian), 1905–1972.
Title: Solid analytic geometry / A. Adrian Albert, Eliakim Hastings Moore
  Distinguished Service Professor of Mathematics, The University of Chicago.
Description: Mineola, New York : Dover Publications, Inc., 2016. | Originally
  published: New York : McGraw-Hill, 1949.
Identifiers: LCCN 2016013941| ISBN 9780486810263 | ISBN 0486810267
Subjects: LCSH: Geometry, Analytic—Solid.

Classification: LCC QA553 .A58 2016 | DDC 516.3/3—dc23 LC record available
at http://lccn.loc.gov/2016013941

Manufactured in the United States by RR Donnelley
81026701   2016
www.doverpublications.com

PREFACE

In a recent text on college algebra the author gave a brief presentation of what seems to him to be the best basis for a modern course on plane analytic geometry, that is, the *algebraic vector* approach. The present text contains an extension of this approach, yielding an exposition in full for the three-dimensional case, and thereby ties up the study of space analytic geometry with the theory of vectors and matrices.

Chapters 1 and 2 contain a treatment of the equations of lines and planes. After a preliminary study of the linear operations on n-dimensional vectors, rectangular coordinates are introduced, three-dimensional vectors and inner products are interpreted geometrically, and, from a consideration of scalar products and axis translations, the parametric equations of a line are obtained. The vector approach then yields a very simple derivation of the normal form of an equation of a plane, and the standard forms of plane and line equations are rather immediate consequences.

Chapter 3 contains an exposition of classical elementary surface and curve theory. Chapter 4 contains the usual treatment of spheres, and Chapter 5 gives the classical descriptions of quadric surfaces in standard position. The latter chapter ends with a rather novel classification of quadrics according to certain invariants.

Chapter 6 is an exposition of that part of the theory of matrices which is needed for a complete development of the so-called *principal axis* transformation. A full account of the orthogonal reduction of a real quadratic form in n variables is given, and the theory is applied in Chapter 7 to the three-dimensional case of quadric surfaces. This latter chapter ends with a discussion of the symmetries of quadric surfaces.

The first seven chapters of this text provide an exposition of the basic topics of solid analytic geometry, the material being adequate for a one-quarter course on the subject. The remaining chapters contain additional material for longer courses or outside reading. Chapter 8, on spherical coordinates. contains a discussion of some practical aspects of the theory of rotations and translations of axes in space. It is quite clear that rectangular coordinates are not as practical for actual measurements as are the coordinates of range, azimuth, and elevation, and in this chapter methods are developed for actual computation of the changes in these measurable coordinates after translations or rotations of axes. The chapter ends with a discussion of gnomonic charts.

Our final chapter contains a brief presentation of the elements of projective geometry. The effect on the theory of linear transformations of the use of homogeneous coordinates is given, and the chapter contains a rigorous matrix proof of the invariance of the cross ratio under projective transformations. It is hoped that the use of modern algebraic techniques in this and in the earlier chapters of the present text will serve to make the subject of solid analytic geometry fit better in the teaching of modern mathematics than it has in the past.

ADRIAN ALBERT

CHICAGO, ILL.
March, 1949

CONTENTS

PREFACE v

1. COORDINATES AND LINES 1
   1. Vectors 1
   2. Scalar multiplication 2
   3. Inner products. 3
   4. The angle between two vectors. 4
   5. Directed lines 5
   6. Orthogonal projections 6
   7. Rectangular coordinates in ordinary space. 7
   8. The length of a vector in space. 10
   9. Lines through the origin. 10
  10. The angle between two vectors in space. 12
  11. Translation of axes 13
  12. Geometric addition of vectors 15
  13. The length of a line segment. 16
  14. Direction numbers 16
  15. Equations of a line 18
  16. Equations in symmetric form. 20
  17. Distance between a line and a point. 23

2. PLANES 24
   1. The normal form. 24
   2. The general equation 25
   3. Planes through three points 27
   4. Parallel planes 27
   5. Distance from a plane to a point 29
   6. Angle between two planes 30
   7. The line of intersection of two planes 30
   8. Angle between a line and a plane. 32
   9. Pencils of planes 32
  10. Parametric equations of a plane 35

3. SURFACES AND CURVES 37
   1. Equations of a surface. 37
   2. Space curves. 38
   3. Plane sections 38
   4. Algebraic surfaces. 40
   5. Cones. 40
   6. Cylinders 43
   7. Surfaces of revolution. 45

   8. Symmetries of surfaces 46
   9. Intersections of a line and a surface 47
  10. Lines on a cylinder 48
  11. Tangent lines and planes 49
  12. Tangents to quadrics 52

4. SPHERES 54
   1. Equations of spheres 54
   2. Spheres satisfying given conditions 55
   3. Linear families of spheres 56
   4. Angles between spheres 58

5. QUADRIC SURFACES 59
   1. Ellipsoids 59
   2. Quadric cones 62
   3. Hyperboloids 63
   4. Lines on a hyperboloid 67
   5. Paraboloids 70
   6. Cylinders 74
   7. Classification of quadric surfaces 75

6. THEORY OF MATRICES 77
   1. Matrices 77
   2. Addition and scalar multiplication 78
   3. Matrix multiplication 79
   4. Transposition 81
   5. Special matrices 82
   6. Products in terms of rows and columns 83
   7. Partitioning of a matrix 84
   8. Determinants 86
   9. Properties of determinants 88
  10. The inverse of a matrix 90
  11. Linear systems of equations 92
  12. Homogeneous systems 94
  13. The characteristic equation 95
  14. Similar matrices 96
  15. Real symmetric matrices 97
  16. Orthogonal matrices 98
  17. Orthogonal reduction of a symmetric matrix 100
  18. Uniqueness of characteristic unit vectors 101

7. ROTATIONS OF AXES AND APPLICATIONS 103
   1. Orthogonal transformations 103
   2. Products of orthogonal transformations 106
   3. Reflections and rotations 106
   4. Orthogonal reduction of a real quadratic form 110
   5. Quadric surfaces 114
   6. Plane sections of quadrics 116

CONTENTS

7. Points of symmetry. 117
8. Planes of symmetry. 118
9. Lines of symmetry 121

8. SPHERICAL COORDINATES 123
   1. Azimuth and elevation 123
   2. The angle between two vectors. 125
   3. Parallax. 127
   4. Other spherical coordinates 131
   5. The matrices of planar rotations 133
   6. Rotations as products of planar rotations 136
   7. Stabilization of coordinates 137
   8. Gnomonic projections. 139
   9. Gnomonic charts. 141

9. ELEMENTS OF PROJECTIVE GEOMETRY. 143
   1. Homogeneous coordinates 143
   2. Lines and planes 143
   3. Projective transformations. 146
   4. Tetrahedral coordinates. 147
   5. The unit point. 149
   6. Invariant points 150
   7. Quadric surfaces 150
   8. Cross ratios of points 152
   9. Plane coordinates and duality 158

INDEX. 159

CHAPTER 1
COORDINATES AND LINES

1. Vectors. A sequence $P = (x_1, \ldots, x_n)$ of n numbers x_i is called an *n-dimensional vector*. The elements $x_1, \ldots, x_n$ are called the *coordinates* of P and x_i is the ith coordinate. We shall limit our attention to *real vectors, i.e.,* to vectors whose coordinates are real numbers.

The vector whose coordinates are all zero is called the *zero* vector and will be designated by 0. A real vector may be interpreted as a representation, relative to a fixed coordinate system with O as origin, of a point in n-dimensional real Euclidean space.

It may also be interpreted as the line segment $\overrightarrow{OP}$ directed from O to P. These interpretations have little intuitive significance except for the cases $n \leq 3$, and we shall carry out the details in this text for the case $n = 3$.

The *sum* $P + Q$ of two vectors $P = (x_1, \ldots, x_n)$ and $Q = (y_1, \ldots, y_n)$ is the vector $(x_1 + y_1, \ldots, x_n + y_n)$ whose ith coordinate is the sum $x_i + y_i$ of the ith coordinate of P and the ith coordinate of Q. We leave the verification of the following simple results to the reader:

Lemma 1. *Addition of vectors is commutative, that is,* $P + Q = Q + P$ *for all vectors* P *and* Q.

Lemma 2. *Addition of vectors is associative, that is,* $(P + Q) + R = P + (Q + R)$ *for all vectors* P, Q, R.

Lemma 3. *The zero vector* 0 *has the property that* $P + 0 = P$ *for all vectors* P.

Lemma 4. *Let* $P = (x_1, \ldots, x_n)$. *Then the vector* $-P = (-x_1, \ldots, -x_n)$ *has the property that* $P + (-P) = 0$.

Lemma 5. *If* P *and* Q *are any vectors the equation* $P + X = Q$ *has the solution* $X = Q + (-P)$. *We call this vector the* **difference** *of* Q *and* P *and write* $X = Q - P$. *Then the* ith *coordinate of* $Q - P$ *is the difference of the* ith *coordinate of* Q *and the* ith *coordinate of* P.

1

EXERCISE

Verify the five lemmas.

2. Scalar multiplication. If a is a number and $P = (x_1, \ldots, x_n)$ is a vector, we define the *scalar product* of P by a to be

$$aP = (ax_1, \ldots, ax_n).$$

Evidently $1P = P$, $(-1)P = -P$, $0P = 0$. The reader should verify that

$$a(bP) = (ab)P, \qquad (a+b)P = aP + bP, \qquad a(P+Q) = aP + aQ$$

for all scalars a and b and all vectors P and Q.

A sum

$$P = a_1P_1 + \cdots + a_mP_m,$$

of scalar products a_jP_j of vectors P_j by scalars a_j, is called a *linear combination* of $P_1, \ldots, P_m$. We shall say that $P_1, \ldots, P_m$ are *linearly independent* vectors if it is true that a linear combination $a_1P_1 + \cdots + a_mP_m = 0$ if and only if $a_1, \ldots, a_m$ are all zero. If $P_1, \ldots, P_m$ are not linearly independent, we shall say that $P_1, \ldots, P_m$ are linearly *dependent*.

Let E_i be the vector whose ith coordinate is 1 and whose other coordinates are all zero. Then

$$P = (x_1, \ldots, x_n) = x_1E_1 + \cdots + x_nE_n.$$

Thus every vector is a linear combination of $E_1, \ldots, E_n$. If $x_1E_1 + \cdots + x_nE_n = P = 0$, then $(x_1, \ldots, x_n) = 0$, that is, $x_1 = x_2 = \cdots = x_n = 0$. It follows that $E_1, \ldots, E_n$ are linearly independent.

EXERCISES

1. Show that if $P = (x_1, \ldots, x_n)$ and $Q = (y_1, \ldots, y_n)$ are not zero then P and Q are linearly dependent if and only if Q is a scalar multiple of P.

2. Show that if $P_1, \ldots, P_m$ are linearly independent and P_{m+1} is another vector then $P_1, \ldots, P_m, P_{m+1}$ are linearly dependent if and only if P_{m+1} is a linear combination of $P_1, \ldots, P_m$.

3. Compute the following linear combinations of $P_1 = (1, -1, 2, 3)$, $P_2 = (0, 1, -1, 2)$, $P_3 = (-2, 1, -1, 2)$.

   (a) $2P_1 + P_2 + P_3$ (b) $P_1 + 3P_2 - 2P_3$ (c) $3P_1 + 2P_2 - 4P_3$

4. Use Exercises 1 and 2 in determining which of the following sets of three vectors are linearly independent sets.

(a) $(1, -1, 2)$, $(1, 1, 0)$, $(0, -1, 1)$

(b) $(2, 1, 1)$, $(1, -1, 1)$, $(5, 4, 2)$

(c) $(1, 0, -2)$, $(2, -1, 2)$, $(4, -3, 10)$

(d) $(1, -1, 1)$, $(-1, 2, 1)$, $(-1, 2, 2)$

(e) $(1, 0, -1, 1)$, $(0, -1, 1, -1)$, $(4, -1, -3, 4)$

(f) $(5, 1, -2, -6)$, $(1, 1, 0, -2)$, $(2, -1, -1, 0)$

(g) $(1, 0, 0, 0)$, $(1, 1, 1, 1)$, $(3, 1, 1, 1)$

(h) $(1, 1, -1, 2)$, $(2, 2, -2, 3)$, $(3, 3, -2, 6)$

5. Prove that *any three* two-dimensional vectors are linearly dependent.

6. Prove that *any four* three-dimensional vectors are linearly dependent.

3. Inner products. If $P = (x_1, \ldots, x_n)$ and $Q = (y_1, \ldots, y_n)$ are any two vectors, we shall call the *number*

$$(1) \qquad P \cdot Q = x_1 y_1 + \cdots + x_n y_n$$

the *inner product* of P and Q. Evidently, $P \cdot Q = Q \cdot P$.

The *norm* of a vector P is defined to be the inner product

$$(2) \qquad P \cdot P = x_1{}^2 + \cdots + x_n{}^2.$$

If P is any real vector, the number $P \cdot P \geqq 0$ and has a nonnegative square root

$$(3) \qquad t = \sqrt{P \cdot P} = \sqrt{x_1{}^2 + \cdots + x_n{}^2},$$

which we shall call the *length* of P.

A vector P is called a *unit* vector if $P \cdot P = 1$. Thus a real unit vector is a vector whose length (and whose norm) is 1.

Lemma 6. *Every real nonzero vector is a scalar multiple of exactly two unit vectors. These are the vectors* $U = t^{-1}P$ *and* $-U$, *where* t *is the length of* P. *Then if* $P = tU$, *where* $t > 0$ *and* U *is a unit vector, the number* t *is the length of* P.

For proof we first let $P = tU$ where $U = (u_1, \ldots, u_n)$ is a unit vector. Then $P \cdot P = (tu_1)^2 + \cdots + (tu_n)^2 = t^2(u_1{}^2 + \cdots + u_t{}^2) = t^2$, and $t = \pm \sqrt{P \cdot P}$; $t = \sqrt{P \cdot P}$, if $t \geqq 0$. Conversely, let $U = t^{-1}P$, where $t = \sqrt{P \cdot P}$. Then $U \cdot U = (t^{-1}x_1)^2 + \cdots + (t^{-1}x_n)^2 = t^{-2}(x_1{}^2 + \cdots + x_n{}^2) = 1$ and U is a unit vector. The vector $-U = -t^{-1}P$ is clearly also a unit vector.

EXERCISES

1. Give the norms and lengths of the following vectors:

(a) $(2, 2, -1)$

(b) $(1, 1, 0)$

(c) $(1, -4, 8)$

(d) $(1, -1, 1, -1)$

(e) $(1, -1, 2, 1)$

(f) $(3, 2, -1, 1, 1)$

2. Give the unit vector $(\sqrt{P \cdot P})^{-1}P$ for each vector of Exercise 1.

4. The angle between two vectors. If $P = (x_1, \ldots, x_n)$ and $Q = (y_1, \ldots, y_n)$ are any two real nonzero vectors, the difference

$$(4) \qquad (P \cdot P)(Q \cdot Q) - (P \cdot Q)^2 \geqq 0.$$

For $(P \cdot P)(Q \cdot Q) = (x_1^2 + \cdots + x_n^2)(y_1^2 + \cdots + y_n^2)$ is the sum of $x_1^2y_1^2 + x_2^2y_2^2 + \cdots + x_n^2y_n^2$ and all expressions of the form $(x_iy_i)^2 + (x_jy_i)^2$ for $1 \leqq i < j \leqq n$. The square $(P \cdot Q)^2 = (x_1y_1 + \cdots + x_ny_n)^2$ is the sum of $x_1^2y_1^2 + x_2^2y_2^2 + \cdots + x_n^2y_n^2$ and all products of the form $2x_iy_ix_jy_j$ for $1 \leqq i < j \leqq n$. The difference then is the sum of all expressions of the form $(x_iy_i)^2 + (x_jy_i)^2 - 2x_iy_ix_jy_j = (x_iy_j - x_jy_i)^2$ for $1 \leqq i < j \leqq n$, and must be nonnegative.

The numbers $P \cdot P$, $Q \cdot Q$, and $(P \cdot Q)^2$ are all positive, and we have shown that

$$0 \leqq \frac{(P \cdot Q)^2}{(P \cdot P)(Q \cdot Q)} \leqq 1.$$

It follows that there exists an angle θ between 0 and 180° such that

$$(5) \qquad \cos \theta = \frac{P \cdot Q}{\sqrt{P \cdot P} \sqrt{Q \cdot Q}}.$$

We define this angle θ to be the *angle between* the vectors P and Q.

Two vectors are said to be *orthogonal* (*i.e.*, perpendicular) if $\cos \theta = 0$. Then P and Q are orthogonal if and only if their inner product

$$(6) \qquad P \cdot Q = x_1y_1 + \cdots + x_ny_n = 0.$$

Thus, if P and Q are any vectors, we multiply corresponding coordinates and add the products. The sum so obtained is zero if and only if P and Q are orthogonal.

EXERCISES

1. Compute $P \cdot Q$ for each of the following vector pairs P, Q:

(a) $(1, 1, -1)$, $(1, 0, 1)$ (f) $(1, -1, 1, 1)$, $(1, 1, 1, 0)$
(b) $(1, 2, 3)$, $(-1, 1, -1)$ (g) $(2, 3, -1, 6)$, $(3, -2, 6, 1)$
(c) $(1, 1, 2)$, $(0, -1, 1)$ (h) $(1, 2, 3, 4)$, $(2, -1, -1, 1)$
(d) $(-1, 0, 1)$, $(2, 1, 1)$ (i) $(4, -6, 1, 2)$, $(1, 2, -1, 2)$
(e) $(-1, 3, 2)$, $(1, 1, -1)$ (j) $(3, 1, -1, 1)$, $(0, 1, 1, 0)$

2. Which pairs are orthogonal?
3. Compute $\cos \theta$ for each nonorthogonal pair.

5. Directed lines. Directed lines are frequently used in the geometry of three-dimensional Euclidean space, *i.e.*, in ordinary solid analytic geometry. Every pair of distinct points P and Q in space determines a line passing through P and Q. We shall use the notation PQ for this line and shall prefix the word *ray* when we mean the ray PQ, which is the *half* line *from* P *through* Q.

Let us assume that a unit of measurement has been prescribed and that we have measured the length of the line segment joining P to Q in terms of this unit. The result is a real number that is positive if P and Q are different points and is zero only when P and Q coincide. We shall use the notation $|PQ|$ for this measurement of length.

When P and Q are points on a directed line, we shall use the symbol $\overline{PQ}$ for the *signed* length of the segment joining P to Q and directed from P toward Q. Then $\overline{PQ} = |PQ|$ if the direction from P to Q is the positive direction on the line, and $\overline{PQ} = -|PQ|$ if the direction from P to Q is opposite to the positive direction on the line. See Fig. 1 in which $\overline{PQ} > 0$ and $\overline{RQ} < 0$, and note that in all cases $\overline{PQ} = -\overline{QP}$.

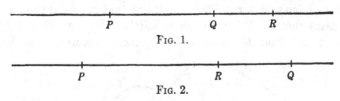

P   Q   R
Fig. 1.

P   R   Q
Fig. 2.

If P, Q, R are on a directed line, it should be clear from Figs. 1 and 2 that $\overline{PQ} + \overline{QR} = \overline{PR}$. This equation may be generalized to the case of any finite number of points on a directed line and

the generalized equation is

(7) $$\overrightarrow{P_1P_2} + \overrightarrow{P_2P_3} + \cdots + \overrightarrow{P_{n-1}P_n} = \overrightarrow{P_1P_n}.$$

6. Orthogonal projections. A theorem of solid geometry states that through a given point P there exists precisely one plane perpendicular to a given line L. This plane intersects L in a point P' such that the line PP' is perpendicular to L. We shall call P' the *orthogonal projection* of P on L.

If P and Q are any two points, we shall designate by $\overrightarrow{PQ}$ the *line segment* which joins P to Q and which is directed from P to Q. Project P and Q orthogonally on a directed line L, and obtain projections P' and Q'. Then we define the orthogonal projection of $\overrightarrow{PQ}$ on L to be the *signed length* $\overline{P'Q'}$. It follows that the orthogonal projection of $\overrightarrow{QP}$ is the negative $\overline{Q'P'}$ of the orthogonal projection of $\overrightarrow{PQ}$.

A directed *broken line* joining two points P and Q is the geometric configuration consisting of the directed line segments $\overrightarrow{PP_1}, \overrightarrow{P_1P_2}, \ldots, \overrightarrow{P_nQ}$ for any finite number n of points $P_1, \ldots, P_n$. Let us use the notation $\overrightarrow{PP_1 \cdots P_nQ}$ for such a configuration and define the orthogonal projection of $\overrightarrow{PP_1 \cdots P_nQ}$ to be the sum $\overline{P'P_1'} + \overline{P_1'P_2'} + \cdots + \overline{P_n'Q'}$. By formula (7) this sum is equal to $P'Q'$ and we have proved the following:

Theorem 1. *The orthogonal projection of any directed line segment* $\overrightarrow{PQ}$ *on a directed line* L *is equal to the orthogonal projection on* L *of any directed broken line from* P *to* Q.

As we have said, a ray PQ is a half line that begins at the point P, passes through Q, and extends indefinitely. If PQ and PR are two rays from the same point P, there is a unique angle θ

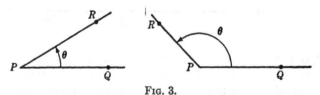

Fig. 3.

between them such that $0 \leqq \theta \leqq 180°$. We define this angle to be the *angle between* the ray PQ and the ray PR.

Let P and Q be any points and P' and Q' be their respective projections on a directed line L, as in Fig. 4. Then Q, P, P' are three of the vertices of a parallelogram, and we find the fourth vertex R by drawing a line segment $P'R$ such that $|P'R| = |PQ|$.

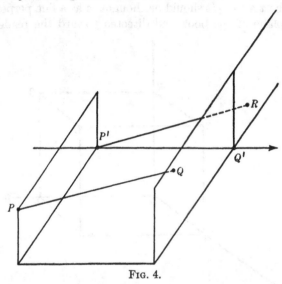

Fɪɢ. 4.

Define the angle θ between the ray PQ and the line L to be the angle between the ray $P'R$ and the ray from P' in the positive direction on L. By the standard ratio definition of the cosine of an angle we have

$$\cos \theta = \frac{\overline{P'Q'}}{|P'R|} = \frac{\overline{P'Q'}}{|PQ|}$$

We have proved the following:

Theorem 2. *Let θ be the angle between $\overrightarrow{PQ}$ and a directed line* L. *Then the projection of $\overrightarrow{PQ}$ on* L *is equal to* $|PQ| \cos \theta$.

7. Rectangular coordinates in ordinary space. A rectangular Cartesian coordinate system in ordinary three-dimensional real Euclidean space is a certain one-to-one correspondence between the points of space and three-dimensional real vectors (x, y, z). The construction of the correspondence begins with the construction of three mutually perpendicular directed lines intersecting at a point O called the *origin* of the coordinate system (see Fig. 5).

The three lines are called *coordinate axes*. The first of them is a vertical line directed upward. It is called the z axis. The second line is a horizontal line in the plane of the book and is directed to the right. It is called the y axis. The remaining line is the x axis. It should be thought of as a line perpendicular to the plane of the book and directed toward the reader. The

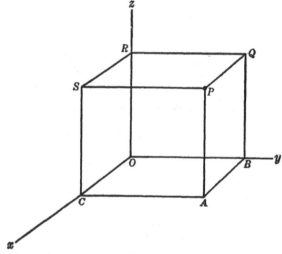

Fig. 5.

specification of a coordinate system will be completed as soon as a unit of measurement, which will be used for *all* measurements of lengths of lines, is given. This is usually done by specifying a unit point U on the x axis such that $\overline{OU}$ is the unit of length.

Let P be any point in space, x be the projection of $\vec{OP}$ on the x axis, y be the projection of $\vec{OP}$ on the y axis, and z be the projection of $\vec{OP}$ on the z axis. Then the vector (x, y, z) is uniquely determined by P and we write $P = (x, y, z)$.

Conversely, if (x, y, z) is given, we can draw a plane perpendicular to the x axis and through a point C on the x axis such that $\overline{OC} = x$. All points P on this plane have the property that the projection of $\vec{OP}$ on the x axis is x. We may construct another plane perpendicular to the y axis through a point B on the y axis such that $\overline{OB} = y$, and finally a plane perpendicular to the z axis and through a point R on the z axis such that $\overline{OR} = z$. The intersection of these three mutually perpendicular planes is the unique point which is such that $P = (x, y, z)$.

This completes our description of a rectangular coordinate system in ordinary space. Let us now observe some simple properties of a coordinate system. It should be clear that each pair of axes defines a plane. We call the three mutually perpendicular planes that are so defined the *coordinate planes*. The x, y plane is that determined by the x and y axes and z is the perpendicular distance from the x, y plane to $P = (x, y, z)$. The y, z and z, x planes are defined similarly, and x and y are also perpendicular distances from coordinate planes to P.

The coordinates x, y, z are any real numbers and therefore may be positive, negative, or zero. The reader should verify the following statements and answer the questions:

1. The x axis is the set of all points P such that $y = z = 0$. What are the corresponding equations for the y axis? The z axis?

2. The x, y plane is the set of all points such that $z = 0$. What is the corresponding equation for the y, z plane? The z, x plane? A plane parallel to the x, y plane and three units above it? A plane parallel to the y, z plane and three units behind it?

3. The coordinate planes divide all of space into octants. The forward, right-hand, upper octant is that where $x \geqq 0$, $y \geqq 0$, $z \geqq 0$. We call this the *first octant*, label it I, and draw most of our diagrams as if the points being studied are all in the first octant. This will restrict only the generality of the diagrams but not that of the mathematical arguments.

4. The octants labeled II to VIII are those where the corresponding sets of signs are given as follows: $(-, +, +)$, $(-, -, +)$, $(+, -, +)$, $(+, +, -)$, $(-, +, -)$, $(-, -, -)$, $(+, -, -)$. Describe the octants in terms of the words *forward* and *backward*, *right* and *left*, and *upper* and *lower*.

EXERCISES

1. Verify the statements, and answer the questions listed above.

2. How can an equation be used to describe the set of all points with equal x and y coordinates? Give the equation and the nature of the geometric configuration.

3. Let a given point P be such that the projection $\overline{OP'}$ of $\overrightarrow{OP}$ on the x axis is equal to $\cos\alpha$, where α is the angle between the positive x axis and OP. What is the projection of $\overrightarrow{OP'}$ on $\overrightarrow{OP}$?

8. The length of a vector in space. A directed line segment $\overrightarrow{OP}$ is usually called a *vector* and this is the inspiration of the name we have given to number sequences (x, y, z). We shall now show that if $P = (x, y, z)$, then the length of OP is given by

$$|OP| = \sqrt{x^2 + y^2 + z^2},$$

that is, our formula for the length of a vector is an actual length in the three-dimensional case.

Use Fig. 5 and observe that $|OA|$ is the diagonal of a rectangle whose sides are x and y. Then $|OA|^2 = x^2 + y^2$. Also, $\overline{AP} = z$ and $|OP|$ is the hypotenuse of a right triangle whose legs are $|OA|$ and $|AP|$. By the theorem of Pythagoras $|OP|^2 = |OA|^2 + z^2 = x^2 + y^2 + z^2$. This yields the formula above.

ORAL EXERCISE

Give the lengths of the following vectors:

(a) $(-1, 2, 3)$	(d) $(-2, 2, 1)$	(g) $(0, 0, -3)$
(b) $(1, 1, 2)$	(e) $(3, 6, 0)$	(h) $(3, 0, -4)$
(c) $(1, -2, -2)$	(f) $(-1, 1, 1)$	(i) $(5, 7, -1)$

9. Lines through the origin. Every line OP through the origin is composed of two rays. One of them is the ray OP and the other is the ray OQ, where $Q = -P$. For if $P = (x, y, z)$ is on a line and $-P = (-x, -y, -z)$, it should be evident geometrically that $-P$ is also on the line.

Every ray OP contains a point U such that $|OU| = 1$. Then $U = (\lambda, \mu, \nu)$ is a unit vector and

$$(8) \qquad\qquad \lambda^2 + \mu^2 + \nu^2 = 1.$$

There is exactly one other unit vector on the line OP and this is the point $-U = (-\lambda, -\mu, -\nu)$.

Define α to be the angle between the positive x axis and the ray OP, β to be the angle between the positive y axis and the ray OP, and γ to be the angle between the positive z axis and the ray OP. By Theorem 2 and our definition of coordinates

$$(9) \quad x = |OP| \cos\alpha, \qquad y = |OP| \cos\beta, \qquad z = |OP| \cos\gamma.$$

We call α, β, γ the *direction angles* of the ray OP, and it should be clear that any two of them and the *sign* of the third uniquely determine this ray. In fact, the set of all rays making an angle with the positive x axis form a cone of lines, the set of all rays making an angle β with the positive y axis form a cone of lines, and the ray OP is one of the two intersections of these two cones. The dependence of γ on α and β is expressed by formula (11) below.

Since the point U is on the ray OP and $|OU| = 1$, we may apply formula (9) to see that

(10) $\lambda = \cos \alpha, \qquad \mu = \cos \beta, \qquad \nu = \cos \gamma.$

By formula (8)

(11) $$\cos^2 \alpha + \cos^2 \beta + \cos^2 \gamma = 1.$$

This relation expresses the dependence of the three angles α, β, γ. The numbers $\cos \alpha$, $\cos \beta$, and $\cos \gamma$ are called the *direction cosines* of the ray OP, and we have seen that they satisfy formula (11) and are the coordinates of the unique unit vector on this ray.

If $Q = (x_1, y_1, z_1)$ is on the ray OP, then $x_1 = |OQ| \cos \alpha$, $y_1 = |OQ| \cos \beta$, $z_1 = |OQ| \cos \gamma$. By formula (9) we have

(12) $$Q = tP, \qquad t = \frac{|OQ|}{|OP|}.$$

If Q is on the line OP but not on the ray OP, then $-Q$ is on the ray OP and $-Q = tP$. We have proved the following geometric interpretation of our algebraically defined operation of scalar product:

Theorem 3. *A point* Q *is on a line through the origin and another point* $P = (x, y, z)$ *if and only if* Q *is a scalar multiple* $t(x, y, z)$ *of* P. *Then* $|t| = |OQ| \cdot |OP|^{-1}$ *and* $t \geqq 0$ *or* $t < 0$ *according as* Q *is or is not on the ray from* O *through* P.

EXERCISES

1. Find the direction cosines of the ray OP for each of the following vectors P:

(a) $(-1, -1, 1)$ (d) $(1, 0, -1)$
(b) $(1, 2, 2)$ (e) $(-1, 1, 2)$
(c) $(-2, 1, 1)$ (f) $(2, -1, 3)$

2. What previous exercise is identical with Exercise 1 except for the actual numbers involved?

3. What are the direction cosines of the coordinate axes?

10. The angle between two vectors in space. Let P and Q be any two points distinct from the origin and θ be the angle between the ray OP and the ray OQ. Let $U = (\lambda, \mu, \nu)$ be the unit vector

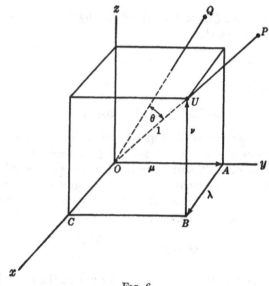

FIG. 6.

on the ray OP and $V = (\lambda_0, \mu_0, \nu_0)$ be the unit vector on the ray OQ. Project $\overrightarrow{OU}$ orthogonally on the line OQ directed from O toward Q. The projection is $|OU| \cos \theta = \cos \theta$. But this is the same as the projection on OQ of the broken line $\overrightarrow{OABU}$ of Fig. 6. The projection of $\overrightarrow{AB}$ is the same as the projection of $\overrightarrow{OC}$ and, since the cosine of the angle between the ray OC and the ray OQ is λ_0, this projection is $\lambda\lambda_0$. Similarly, the projection of $\overrightarrow{OA}$ on OQ is $\mu\mu_0$ and that of $\overrightarrow{BU}$ on OQ is $\nu\nu_0$. Then

(13) $\cos \theta = \lambda\lambda_0 + \mu\mu_0 + \nu\nu_0.$

If $P = (x, y, z)$ and $Q = (x_0, y_0, z_0)$, then formula (13) becomes

$$(14) \qquad \cos \theta = \frac{xx_0 + yy_0 + zz_0}{\sqrt{x^2 + y^2 + z^2} \sqrt{x_0^2 + y_0^2 + z_0^2}}.$$

For $|OP| = \sqrt{x^2 + y^2 + z^2}$, $|OQ| = \sqrt{x_0^2 + y_0^2 + z_0^2}$, and $x = |OP|\lambda$, $y = |OP|\mu$, $z = |OP|\nu$, $x_0 = |OQ|\lambda_0$, $y_0 = |OQ|\mu_0$, $z_0 = |OQ|\nu_0$. We also see that a line through O and $P = (x, y, z)$ and a line through 0 and $Q = (x_0, y_0, z_0)$ are perpendicular if and only if $xx_0 + yz_0 + zz_0 = 0$.

The exercises of Sec. 4 are exercises on the material of this section.

ORAL EXERCISE

Show that the triangles whose vertices are P, Q, and the origin are right triangles in the following cases:

(a) $P = (2, -1, 4)$, $Q = (3, 2, -1)$
(b) $P = (1, -1, 0)$, $Q = (1, 1, 6)$
(c) $P = (2, 1, -2)$, $Q = (3, 4, 5)$
(d) $P = (1, 1, -2)$, $Q = (1, 1, 1)$

11. Translation of axes. The correspondence between points P and vectors x, y, z depends on the use of a fixed coordinate system. If the coordinate system is altered, so is the correspondence.

It is desirable to investigate the effect of a *translation* of axes on this correspondence. Such a change is the result of setting up an x', y', z' coordinate system in which the (new) x' axis is parallel to the x axis, the y' axis is parallel to the y axis, and the z' axis is parallel to the z axis. Thus the x', y', z' coordinate system may be conceived of as having been obtained by a motion called a *translation* in which axes are moved parallel to themselves and the main effect is a change in origin.

Every point P in space will now have two sets of coordinates. We will call the first set the x, y, z coordinates and the second set the x', y', z' coordinates. The origin O' of the second coordinate system has a set of x, y, z coordinates that we shall designate by x_0, y_0, z_0, as in Fig. 7.

Suppose now that P has coordinates x, y, z. This means that P is z units *above the x, y plane*. Note that z units above means $-z$ units below if z is negative. The x', y' plane is z_0 units above the x, y plane and it should be clear then that P is only $z - z_0$ units above the x', y' plane, that is, $z' = z - z_0$. By similar considerations we obtain the relations

(15) $x' = x - x_0,$ $y' = y - y_0,$ $z' = z - z_0$

between the two sets of coordinates of P.

Conversely, if x', y', z' and x_0, y_0, z_0 are given, we can use the relations

(16) $x = x' + x_0,$ $y = y' + y_0,$ $z = z' + z_0,$

between the two sets of coordinates.

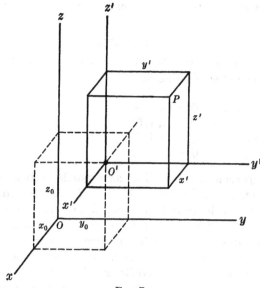

FIG. 7.

Formula (16) is sometimes used in making a translation of axes, which simplifies what we shall later call an equation of a surface.

EXERCISES

1. What translation of axes will simplify the following equations of surfaces?

(a) $\dfrac{(x-1)^2}{4} + \dfrac{(y+2)^2}{6} - \dfrac{(z-3)^2}{9} = 1$

(b) $2(x+2)^2 - (y-3)^2 + (z+1)^2 = 5$

(c) $x^2 - 2x + 3y^2 - 6y + 4z^2 - 12z = 0$

(d) $2x^2 + 3y^2 - z^2 + 6x - 12y + z = 9$

(e) $3x^2 - 3y^2 + 6z^2 = 6x - 9y + 24z$

(f) $3x - 2y^2 + z^2 = 6$

2. Find the x', y', z' coordinates of each of the following points P if there is a translation of axes with origin at O'.

(a) $P = (1, -1, 1)$, $O' = (2, 2, -1)$
(b) $P = (0, 0, 0)$, $O' = (1, -1, 1)$
(c) $P = (-1, -1, 1)$, $O' = (-1, -1, 2)$
(d) $P = (3, 1, 1)$, $O' = (-2, 2, 1)$
(e) $P = (1, 2, -3)$, $O' = (1, 0, -1)$
(f) $P = (2, 4, 0)$, $O' = (0, -1, 0)$

3. Let the vectors P in Exercise 2 be the x', y', z' coordinates of corresponding points. Give their x, y, z coordinates.

4. Find the x', y', z' equation of the surfaces given by the equations of Exercise 1 after a translation of axes moving the origin to $O' = (1, -1, 2)$.

12. Geometric addition of vectors. Let $P_1 = (x_1, y_1, z_1)$ and $P_2 = (x_2, y_2, z_2)$ be two vectors. Then P_1OP_2 determines a plane

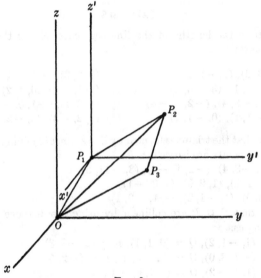

FIG. 8.

and, as in Fig. 8, we can determine a point P_3 in this plane such that OP_3 is parallel to P_1P_2 and in a corresponding direction, $|OP_3| = |P_1P_2|$. The coordinates of P_2, relative to an x', y', z' coordinate system with origin at P_1, will be $(x_2 - x_1, y_2 - y_1, z_2 - z_1)$. But P_3 is located with respect to the x, y, z axes exactly where P_2 is located with respect to the x', y', z' axes. It follows that

$$(17) \qquad P_3 = P_2 - P_1 = (x_2 - x_1, y_2 - y_1, z_2 - z_1).$$

This shows that to construct the vector difference $P_3 = P_2 - P_1$ we construct OP_3 parallel to P_1P_2 and having the same length.

Since $P_3 = P_2 - P_1$, we have $P_2 = P_3 + P_1$. This shows that if P_1 and P_3 are given, their sum, $P_1 + P_3$, is the end point of the diagonal of the plane parallelogram determined by the directed line segments OP_1 and OP_3.

13. The length of a line segment. If $P_1 = (x_1, y_1, z_1)$ and $P_2 = (x_2, y_2, z_2)$ are points in space, the line segment P_1P_2 has the same length $|P_1P_2|$ as OP_3, where $P_3 = P_2 - P_1$ (or $P_3 = P_1 - P_2$). It follows that

$$(18) \qquad |P_1P_2| = \sqrt{(x_1 - x_2)^2 + (y_1 - y_2)^2 + (z_1 - z_2)^2}.$$

EXERCISES

1. Compute the lengths of the line segments joining the following pairs of points:

(a) $(0, 0, 0)$, $(2, -2, 1)$ (e) $(1, 2, 3)$, $(-1, -1, 1)$
(b) $(1, -1, 1)$, $(3, -3, 2)$ (f) $(-1, 1, 1)$, $(0, 1, 2)$
(c) $(-1, -3, 4)$, $(-2, 1, -4)$ (g) $(1, 2, 1)$, $(-1, 0, -2)$
(d) $(1, -1, 1)$, $(0, -1, 1)$ (h) $(1, 2, -3)$, $(3, -2, 1)$

2. Prove that the triangles formed by line segments joining the following triples of points are isosceles:

(a) $(-1, -3, 4)$, $(-2, 1, -4)$, $(3, -11, 5)$
(b) $(2, -1, 2)$, $(1, 2, 0)$, $(4, 0, -1)$
(c) $(0, 0, 0)$, $(1, -1, 2)$, $(-1, -2, 1)$

3. Show that P, Q, R are collinear by using the distance formula in the following cases:

(a) $P = (1, -1, 2)$, $Q = (0, 1, 1)$, $R = (2, -3, 3)$
(b) $P = (-1, 3, 0)$, $Q = (3, -5, 4)$, $R = (-2, 5, -1)$
(c) $P = (1, 1, -2)$, $Q = (-1, 0, -4)$, $R = (5, 3, 2)$

14. Direction numbers. If L is any line in space, we may construct a line L' through O parallel to L. The direction in space of L' is determined by that of L, and conversely. But the direction in space of L' is determined by any nonzero vector a, b, c on L'. We shall therefore call any such vector a set of *direction numbers* of L.

If $P_1 = (x_1, y_1, z_1)$ and $P_2 = (x_2, y_2, z_2)$ are on L, then $P_1 - P_2$

$= (x_1 - x_2, y_1 - y_2, z_1 - z_2)$ is on L' and thus the vector $P_1 - P_2$ is a set of direction numbers of L. Conversely, if $P_1 = (x_1, y_1, z_1)$ is on L and a, b, c is a set of direction numbers of L, then we may take $(a, b, c) = P_2 - P_1$, where $P_2 = (x_1 + a, y_1 + b, z_1 + c)$ is a point on L.

Every line L in space may now be prescribed by a point P_1 on L and either a second point P_2 or a set of direction numbers a, b, c. We form

$$\lambda = \frac{a}{\sqrt{a^2 + b^2 + c^2}}, \qquad \mu = \frac{b}{\sqrt{a^2 + b^2 + c^2}},$$

$$\nu = \frac{c}{\sqrt{a^2 + b^2 + c^2}}$$

and λ, μ, ν are the direction cosines of a ray from O parallel to L. We shall call λ, μ, ν a set of direction cosines of L. Then $-\lambda, -\mu, -\nu$ are also a set of direction cosines of L.

Let L_1 and L_2 be any two lines. Then the corresponding lines L_1' and L_2' through O intersect at O and define a single acute angle θ, which we shall define to be the *angle between* L_1 and L_2. Let a_1, b_1, c_1 be a set of direction numbers of L_1 and a_2, b_2, c_2 direction numbers of L_2. Then we may surely use the formula of Sec. 10 and see that

$$\cos \theta = \frac{a_1 a_2 + b_1 b_2 + c_1 c_2}{\sqrt{a_1^2 + b_1^2 + c_1^2} \sqrt{a_2^2 + b_2^2 + c_2^2}}$$

Moreover, two lines are perpendicular if and only if $a_1 a_2 + b_1 b_2 + c_1 c_2 = 0$.

EXERCISES

1. Find a set of direction numbers for each of the lines defined by the pairs of points of Exercise 1 of Sec. 13, and give a set of direction cosines in each case.

2. In each of the following cases the first vector is a point P_1 on a line and the second vector is a set of direction numbers of the line. Give a corresponding point P_2 on the line.

(a) $(1, -1, 2), (3, 2, -1)$ (c) $(2, 4, -1), (-4, -2, 1)$

(b) $(1, 1, 2), (-1, -1, 3)$ (d) $(1, -2, -1), (3, -1, 2)$

3. Show whether or not the line $P_1 P_2$ is perpendicular to the line $P_3 P_4$ in each of the following cases:

(a) $P_1 = (1, -1, 2), P_2 = (3, -2, 1), P_3 = (2, -1, 3), P_4 = (1, 1, 3)$
(b) $P_1 = (3, 2, -1), P_2 = (1, 2, -1), P_3 = (3, 2, -1), P_4 = (1, 3, 0)$
(c) $P_1 = (2, 4, -1), P_2 = (2, 3, 1), P_3 = (4, -1, 2), P_4 = (6, 3, 3)$
(d) $P_1 = (1, 1, 2), P_2 = (-1, -1, 1), P_3 = (1, 1, -1),$
$P_4 = (2, -1, 1)$
(e) $P_1 = (3, -1, 1), P_2 = (1, 1, -2), P_3 = (1, 2, -3),$
$P_4 = (5, 3, -1)$
(f) $P_1 = (2, -1, 0), P_2 = (1, 1, 1), P_3 = (1, 0, 1), P_4 = (-1, 1, 1)$
(g) $P_1 = (1, 2, -1), P_2 = (-1, 2, 1), P_3 = (0, -1, -2),$
$P_4 = (2, -1, -1)$

4. Compute the cosine of the angle between P_1P_2 and P_3P_4 in each of the cases of Exercise 3 where P_1P_2 is not perpendicular to P_3P_4.

5. Show that the triangles whose vertices are P, Q, R are right triangles in the following cases, and compute the cosines of the acute angles.
(a) $P = (1, -2, 3), Q = (3, -3, 7), R = (4, 0, 2)$
(b) $P = (1, -1, 0), Q = (4, 3, 5), R = (3, 0, -2)$
(c) $P = (0, 2, 1), Q = (0, 0, 1), R = (-1, 1, 1)$
(d) $P = (1, 0, -1), Q = (2, 1, 0), R = (3, -1, -2)$

6. Show that the points P, Q, R, S are the vertices of parallelograms, and determine which lines are the pairs of parallel sides.
(a) $P = (2, 6, 3), Q = (3, 2, 2), R = (0, 5, 4), S = (1, 1, 3)$
(b) $P = (2, -2, 4), Q = (1, 0, 2), R = (2, -1, 3), S = (1, 1, 1)$

7. Compute the lengths of the sides of the parallelograms in Exercise 5.

8. Show that the points of Exercise 3 of Sec. 13 are collinear by using direction rather than distance.

15. Equations of a line. If two distinct points $P_1 = (x_1, y_1, z_1)$ and $P_2 = (x_2, y_2, z_2)$ are given, we may translate our coordinate axes to P_1, and the x', y', z' coordinates of P_2 will be $(x_2 - x_1, y_2 - y_1, z_2 - z_1)$. The x', y', z' coordinates of $P = (x, y, z)$ are $x - x_1, y - y_1, z - z_1$, and Theorem 3 implies that P is on P_1P_2 if and only if

$$(19) \quad x - x_1 = t(x_2 - x_1), \quad y - y_1 = t(y_2 - y_1),$$
$$z - z_1 = t(z_2 - z_1),$$

where if we direct the line P_1P_2 from P_1 toward P_2 then

$$(20) \qquad\qquad t = \frac{\overline{P_1P}}{\overline{P_1P_2}}.$$

The three equations of formula (19) are called a set of *parametric equations of the line* through P_1 and P_2. They have the property that if we prescribe the parameter t we obtain the coordinates

x, y, z of a point P on the line which actually has the property (20).   Conversely, if P is a point on the line, its coordinates will satisfy these equations when t is defined by formula (20).

Equations (19) also permit us to find the coordinates of points on P_1P_2 such that P divides the line segment P_1P_2 in a *prescribed ratio*. For the midpoint of P_1P_2 we have $t = \frac{1}{2}$ and thus $x = x_1 + \frac{1}{2}(x_2 - x_1) = \frac{1}{2}(x_2 - x_1)$.   Indeed the midpoint is given by the formulas

$$(21) \qquad x = \frac{x_1 + x_2}{2}, \qquad y = \frac{y_1 + y_2}{2}, \qquad z = \frac{z_1 + z_2}{2}.$$

ILLUSTRATIVE EXAMPLES

I. Find the midpoint P of the line segment P_1P_2 if $P_1 = (6, 3, -4)$, $P_2 = (-2, -1, 2)$.

Solution

The coordinates of P are

$$x = \frac{6 - 2}{2} = 2, \qquad y = \frac{3 - 1}{2} = 1, \qquad z = \frac{-4 + 2}{2} = -1.$$

II. Let the line P_1P_2 of Example I be directed from P_1 toward P_2, and define t by formula (20).   Find the points $P(t)$ corresponding to $t = \frac{1}{3}$, -1, $-\frac{1}{2}$, and 2, and give a rough sketch showing the relative positions of $P_1, P_2, P(\frac{1}{3}), P(-1), P(-\frac{1}{2})$, and $P(2)$.

Solution

We use the formula $P(t) = P_1 + t(P_2 - P_1) = tP_2 + (1 - t)P_1$ and so obtain

$$P(\tfrac{1}{3}) = \tfrac{1}{3}(-2, -1, 2) + \tfrac{2}{3}(6, 3, -4) = (\tfrac{10}{3}, \tfrac{5}{3}, 2),$$
$$P(-1) = -(-2, -1, 2) + 2(6, 3, -4) = (14, 7, -10),$$
$$P(-\tfrac{1}{2}) = -\tfrac{1}{2}(-2, -1, 2) + \tfrac{3}{2}(6, 3, -4) = (10, 5, -7),$$
$$P(2) = 2(-2, -1, 2) - (6, 3, -4) = (-10, -5, 8).$$

The following is a sketch showing the relative positions of the six points.

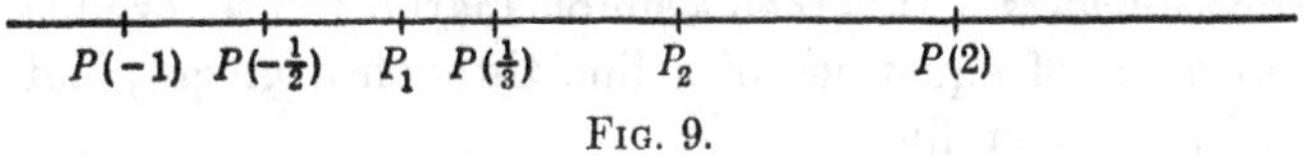

$$P(-1) \quad P(-\tfrac{1}{2}) \quad P_1 \quad P(\tfrac{1}{3}) \qquad P_2 \qquad\qquad P(2)$$

Fig. 9.

EXERCISES

1. Give the midpoints of each of the following line segments P_1P_2:

(a) $(1, -1, 1)$, $(3, -3, 2)$

(b) $(1, -1, 1)$, $(0, 1, -1)$

(c) $(-2, 1, 1)$, $(3, -2, 2)$

(d) $(3, -1, 2)$, $(1, -1, -2)$

(e) $(1, 2, -3)$, $(-3, -2, 1)$

(f) $(4, 1, 6)$, $(2, -1, 4)$

(g) $(1, 0, 2)$, $(2, -1, 1)$

(h) $(1, 1, 1)$, $(2, 2, 5)$

2. Give a set of parametric equations of P_1P_2 in each of the cases of Exercise 1.

3. Work out the problem of Illustrative Example II in each of the cases of Exercise 1 for the values

$$(a) \quad t = \tfrac{2}{3}, \tfrac{3}{2}, -2, -\tfrac{3}{4} \qquad\qquad (b) \quad t = -\tfrac{1}{3}, -2, \tfrac{3}{4}, \tfrac{5}{4}$$

4. The origin and the points P_1, P_2 of Exercise 1 form a triangle in each case. Find the remaining midpoints of sides, and give sets of equations in parametric form of the medians.

16. Equations in symmetric form. If a, b, c is any set of direction numbers of a line through a point $P_1 = (x_1, y_1, z_1)$, then $P_2 = (x_1 + a, y_1 + b, z_1 + c)$ is on the line and formula (19) becomes

$$(22) \qquad x - x_1 = ta, \qquad y - y_1 = tb, \qquad z - z_1 = tc.$$

When $abc \neq 0$, we may write these equations as

$$(23) \qquad \frac{x - x_1}{a} = \frac{y - y_1}{b} = \frac{z - z_1}{c},$$

the common value of the three ratios being t. These new equations are called a set of equations of the given line in *symmetric form*. A special case of this symmetric form is that of

$$(24) \qquad \frac{x - x_1}{\lambda} = \frac{y - y_1}{\mu} = \frac{z - z_1}{\nu},$$

where λ, μ, ν are a set of direction cosines of the given line. If P_1 is given, the form given by (24) is unique apart from the replacement when desired of λ, μ, ν by $-\lambda, -\mu, -\nu$.

Two lines are parallel if and only if they have the same sets of direction numbers. It is then a simple matter to use formula (23) to find a set of equations of a line through a given point and parallel to a given line.

If one of a, b, c is zero in formula (22), we can still use formula (23), provided that we agree that whenever a denominator in formula (23) is zero we shall delete the formal ratio and set the numerator equal to zero; for example, the equations

$$(25) \qquad \frac{x - x_1}{0} = \frac{y - y_1}{b} = \frac{z - z_1}{c}$$

for $bc \neq 0$ shall mean

$$(26) \qquad x = x_1, \qquad \frac{y - y_1}{b} = \frac{z - z_1}{c},$$

and the equations

$$(27) \qquad \frac{x - x_1}{0} = \frac{y - y_1}{0} = \frac{z - z_1}{c}$$

shall mean

$$(28) \qquad x = x_1, \qquad y = y_1.$$

These results are derived from our original equations of formula (22). We may then treat all problems by the use of formula (23) and shall convert formula (25) into formula (26) and formula (27) into formula (28) when the sets of equations assume these irregular forms.

ILLUSTRATIVE EXAMPLES

1. Give a set of equations of the line joining $(3, -1, 2)$ to $(4, 1, -1)$.

Solution

$$\frac{x - 3}{-1} = \frac{y + 1}{-2} = \frac{z - 2}{3}$$

II. Give a set of equations of the line through $(3, -1, 2)$ parallel to the line joining $(1, 2, -4)$ to $(4, -1, -2)$.

Solution

$$\frac{x - 3}{3} = \frac{y + 1}{-3} = \frac{z - 2}{2}$$

III. Give a set of equations of the line joining $(1, 2, 4)$ to $(-1, 3, 4)$.

Solution

The formal set of equations is

$$\frac{x - 1}{2} = \frac{y - 2}{-1} = \frac{z - 4}{0}$$

and these equations must be replaced by

$$\frac{x - 1}{2} = \frac{y - 1}{-1}, \qquad z = 4.$$

IV. Show that the four points $P_1 = (1, -1, 2)$, $P_2 = (2, 1, 3)$,

$P_3 = (4, -1, -2)$, and $P_4 = (5, 1, -1)$ are the vertices of a parallelo-gram, and indicate which lines are the parallel sides.

REMARK: A translation of axes to P_1 replaces our vectors by $P_1 - P_1$, $P_2 - P_1$, $P_3 - P_1$, $P_4 - P_1$. Then $P_2 - P_1$, $P_3 - P_1$, $P_4 - P_1$ must be nonzero vectors, one of these vectors must be the sum of the remain-ing two and these latter two vectors must not have proportional coordi-nates. We have evidently used the parallelogram law for addition of vectors.

Solution

Compute $P_2 - P_1 = (1, 2, 1)$, $P_3 - P_1 = (3, 0, -4)$, and $P_4 - P_1 = (4, 2, -3) = (P_2 - P_1) + (P_3 - P_1)$. This shows that P_1, P_2, P_3, P_4 are the vertices of a parallelogram in which P_1P_2 is parallel to P_3P_4 and P_1P_3 to P_2P_4.

EXERCISES

1. Write a set of equations in parametric form for each of the lines joining the following pairs of points, using formula (23) with the first point given as P_1.

(a) $(2, -1, 3), (3, 1, -2)$ (f) $(1, -1, -2), (2, -2, -3)$
(b) $(2, 1, 2), (2, -1, -2)$ (g) $(2, 1, 3), (2, -1, 2)$
(c) $(3, -1, 2), (-1, 1, 1)$ (h) $(4, 1, 0), (-1, 1, 2)$
(d) $(0, 1, -1), (1, 1, -1)$ (i) $(1, 2, -3), (2, 2, -4)$
(e) $(1, 2, 3), (3, 2, 1)$ (j) $(1, 4, -2), (-1, 3, -1)$

2. Write a set of equations of a line through the origin parallel to the line P_1P_2 for each case of Exercise 1.

3. Write a set of equations of a line through P_1 parallel to the line joining the origin to P_2 for each case of Exercise 1.

4. Write a set of equations of a line through $(-1, 1, 2)$ parallel to the line P_1P_2 for each case of Exercise 1.

5. Show that the following sets of four points are sets of vertices of parallelograms:

(a) $(2, -1, 3), (3, -2, 5), (4, 0, 2), (3, 1, 0)$
(b) $(1, -1, 2), (2, 3, -4), (2, 1, -1), (1, 1, -1)$
(c) $(1, 2, -3), (1, 2, -1), (2, -1, 3), (2, -1, 5)$
(d) $(0, 1, -2), (1, 3, -3), (3, 6, 1), (2, 4, 2)$
(e) $(3, 2, -2), (4, 0, 9), (5, 1, 2), (2, 1, 5)$

6. Write a set of equations for each of the sides of the parallelograms of Exercise 5.

7. Write a set of equations for each of the diagonals.

8. Write a set of equations for each of the lines parallel to a pair of sides and bisecting the other pair of sides.

9. Compute the lengths of the sides and diagonals of the parallelograms of Exercise 5.

17. Distance between a line and a point. We require a procedure for computing the unsigned distance between a line and a point not on the line. Let the line L be given by a set of equations

$$\frac{x - x_1}{\lambda} = \frac{y - y_1}{\mu} = \frac{z - z_1}{\nu},$$

and the point be $P_2 = (x_2, y_2, z_2)$. Then we require $|PP_2|$ where P is the foot of the perpendicular from P_2 to L. If θ is the angle between P_2P_1 and L, then $|PP_2| = |P_2P_1| \cdot |\sin \theta|$ and thus

$$|PP_2|^2 = |P_2P_1|^2(\sin^2 \theta) = [(x_1 - x_2)^2 + (y_1 - y_2)^2 + (z_1 - z_2)^2]$$
$$(1 - \cos^2 \theta).$$

But

$$\cos \theta = \frac{1}{|P_2P_1|} [\lambda(x_2 - x_1) + \mu(y_2 - y_1) + \nu(z_2 - z_1)].$$

It follows that $|PP_2|$ may be computed by the use of

$$(29) \quad |PP_2|^2 = (x_1 - x_2)^2 + (y_1 - y_2)^2 + (z_1 - z_2)^2$$
$$- [\lambda(x_1 - x_2) + \mu(y_1 - y_2) + \nu(z_1 - z_2)]^2.$$

If direction numbers are given rather than direction cosines, it is necessary to compute direction cosines before formula (29) can be applied.

EXERCISES

Compute the distances between points and lines in the following cases:

(a) $P_2 = (0, 0, 0); \dfrac{x - 1}{2} = \dfrac{y}{-1} = \dfrac{z + 2}{3}$

(b) $P_2 = (1, 0, 0); \dfrac{x - 2}{3} = \dfrac{y - 1}{4} = \dfrac{z}{-1}$

(c) $P_2 = (1, -1, 0); \dfrac{x + 1}{-1} = \dfrac{y - 1}{2} = \dfrac{z}{2}$

(d) $P_2 = (0, 1, 2); 3x = 2y = 6z + 12$

(e) $P_2 = (1, 1, -1); 2x = y - 1 = 2z - 4$

(f) $P_2 = (-1, 1, 1); \dfrac{x - 1}{2} = \dfrac{y + 1}{3} = \dfrac{z - 2}{-1}$

(g) $P_2 = (6, 7, 2); \dfrac{x + 2}{4} = \dfrac{y - 1}{3} = \dfrac{z + 2}{2}$

CHAPTER 2

PLANES

1. The normal form. If we draw a ray from the origin O perpendicular to a given plane, the ray is called the ray *normal* to the plane and the line of which it is a part the *normal line*. The ray will have direction cosines λ, μ, ν, which are uniquely determined unless the plane passes through the origin. But in this case the selection of λ, μ, ν, rather than $-\lambda$, $-\mu$, $-\nu$, will not affect our results.

Let p be the distance from 0 to the given plane so that $p \geqq 0$. Then the normal will intersect the plane in the point $P_0 = (p\lambda, p\mu, p\nu)$. A translation of axes, which moves the origin to P_0, replaces the coordinates x, y, z of the arbitrary point P by its x', y', z' coordinates

$$(x - p\lambda, \qquad y - p\mu, \qquad z - p\nu).$$

The ray through P_0 and the vector whose transformed coordinates are (λ, μ, ν) is normal to the given plane. Then P is on the plane if and only if the vector just defined is orthogonal to P_0P, that is, if and only if $(\lambda, \mu, \nu) \cdot (x - p\lambda, y - p\mu, z - p\nu) = 0$. We compute this inner product and see that P is on the given plane if and only if $\lambda x + \mu y + \nu z - p(\lambda^2 + \mu^2 + \nu^2) = 0$. Since $\lambda^2 + \mu^2 + \nu^2 = 1$, we have proved that $P = (x, y, z)$ is on the given plane if and only if

(1) $$\lambda x + \mu y + \nu z = p.$$

We shall call formula (1) *the equation of a plane in normal form*. It is completely unique when $p > 0$. When $p = 0$, the equation $-\lambda x - \mu y - \nu z = 0$ is equally valid and the selection of one set of direction cosines rather than their negatives may be regarded as implying that a positive direction on the line normal to the plane has been selected. This is of little importance.

EXERCISE

Give the equation in normal form of the *two* planes p units from the origin and normal to the line joining P_1 to P_2 in the following cases:

(a) $p = 5$, $P_1 = (-1, 2, 3)$, $P_2 = (1, 2, -1)$
(b) $p = 2$, $P_1 = (-2, 1, -1)$, $P_2 = (1, 3, -2)$
(c) $p = 1$, $P_1 = (0, 0, 0)$, $P_2 = (2, 2, 1)$
(d) $p = \frac{1}{2}$, $P_1 = (2, 1, 4)$, $P_2 = (0, -1, 2)$
(e) $p = 7$, $P_1 = (4, -1, 6)$, $P_2 = (6, -3, 8)$

2. The general equation.   Every linear equation

(2) $$ax + by + cz + d = 0,$$

in which a, b, c, d are real numbers and a, b, c are not all zero, is an equation of a plane; for we can write this equation in an equivalent form

(3) $$\epsilon ax + \epsilon by + \epsilon cz = |d|,$$

where $\epsilon = 1$ if $d \leqq 0$ and $\epsilon = -1$ if $d > 0$.   The number

(4) $$t = \sqrt{a^2 + b^2 + c^2} > 0,$$

and we may divide formula (3) by t to obtain

(5) $$\frac{\epsilon a}{t} x + \frac{\epsilon b}{t} y + \frac{\epsilon c}{t} z = \frac{|d|}{t}.$$

A point $P = (x, y, z)$ satisfies the equation of formula (2) if and only if it satisfies the equation of formula (5).   But then formula (2) is an equation of the plane for which

(6) $$p = \frac{|d|}{t}, \qquad \lambda = \frac{\epsilon a}{t}, \qquad \mu = \frac{\epsilon b}{t}, \qquad \nu = \frac{\epsilon c}{t},$$

i.e., the plane p units from the origin and having a normal ray whose direction cosines are the numbers λ, μ, ν of formula (6). It is important to observe that the original coefficients a, b, c are a set of direction numbers of the normal line.

If $abcd \neq 0$, the general equation may be written in the form

(7) $$\frac{x}{e} + \frac{y}{f} + \frac{z}{g} = 1.$$

This equation is called the *equation of a plane* in *intercept form* and the numbers

(8) $$e = \frac{-d}{a}, \qquad f = \frac{-d}{b}, \qquad g = \frac{-d}{c}$$

are called the *intercepts* of the given plane.   They are indeed the nonzero coordinates of the three intercept points

$$(e, 0, 0), \qquad (0, f, 0), \qquad (0, 0, g)$$

where the given plane cuts the coordinates axes.

If a plane defined by formula (2) passes through a point $P_1 = (x_1, y_1, z_1)$, then $ax_1 + by_1 + cz_1 + d = 0$ and

(9) $$a(x - x_1) + b(y - y_1) + c(z - z_1) = 0.$$

This equation involves the coordinates of P_1 and a set of direction numbers of the normal line. We shall call it the *point, direction number* form of an equation of a plane.

EXERCISES

1. Use formula (9) to write an equation of a plane through P_1 perpendicular to the line joining P_2 to P_3 in each of the following cases:

(a) $P_1 = (-1, 2, 3)$, $P_2 = (-3, 1, 2)$, $P_3 = (-5, 4, 6)$

Ans. $2x - 3y - 4z + 20 = 0.$

(b) $P_1 = (1, -2, -1)$, $P_2 = (4, -1, 2)$, $P_3 = (-1, 4, -1)$
(c) $P_1 = (2, -1, 4)$, $P_2 = (-3, 1, -2)$, $P_3 = (1, 2, -4)$
(d) $P_1 = (-1, 2, -3)$, $P_2 = (-1, 4, 3)$, $P_3 = (0, -3, -4)$
(e) $P_1 = (-2, 1, 3)$, $P_2 = (4, -6, 7)$, $P_3 = (3, -4, 5)$
(f) $P_1 = (1, 4, -1)$, $P_2 = (6, 7, 8)$, $P_4 = (9, 10, 12)$
(g) $P_1 = (3, 2, -2)$, $P_2 = (-9, -8, 0)$, $P_4 = (6, -8, 7)$

2. Find p, λ, μ, ν for each of the following planes:

(a) $2x - 2y + z = 6$ (e) $6y + 2x - 3z = 28$
(b) $3x - 6y + 7z = -1$ (f) $12x + 4y + 6z = -49$
(c) $x + 2y + 2z = -6$ (g) $x + y - 2z = 6$
(d) $-3x + 6y + 2z = 14$ (h) $-x + y + 2z = -3$

3. Find the intercepts of each of the planes in Exercise 2.

4. Find a set of values of λ, μ, ν for each of the following planes through the origin:

(a) $x + y = 0$ (d) $2x + y = 2z$
(b) $3x + 4y = 0$ (e) $-x + y + 2z = 0$
(c) $3x - 4y = 5z$ (f) $3x + 6y + 2z = 0$

5. Write an equation of each of two planes parallel to a corresponding plane of Exercise 4 and one unit from the origin.

6. Write a set of equations in symmetric form of a line perpendicular to the given plane and through the point $(-1, 2, 3)$ in each of the cases of Exercise 2.

3. Planes through three points. Let $P_1 = (x_1, y_1, z_1)$, $P_2 = (x_2, y_2, z_2)$, and $P_3 = (x_3, y_3, z_3)$ be three distinct points. By Sec. 9 of Chap. 1 the three points are collinear if and only if $(x_3 - x_1, y_3 - y_1, z_3 - z_1)$ is a scalar multiple of $(x_2 - x_1, y_2 - y_1, z_2 - z_1)$.

If a plane passes through P_1, its equation is $a(x - x_1) + b(y - y_1) + c(z - z_1) = 0$. If it also passes through P_2 and P_3, we have the relations

(10)
$$a(x_2 - x_1) + b(y_2 - y_1) + c(z_2 - z_1) = 0,$$
$$a(x_3 - x_1) + b(y_3 - y_1) + c(z_3 - z_1) = 0.$$

These two equations in a, b, c do not have proportional coefficients when P_1, P_2, P_3 are not collinear, and they can be solved to yield a set of direction numbers $(a, b, c) = (0, 0, 0)$ of the normal to the given plane. Then formula (9) is an equation of the required plane.

ILLUSTRATIVE EXAMPLE

Find an equation of the plane through $(-1, 2, 3)$, $(-3, 1, 2)$, $(-5, 4, 6)$.

Solution

Equations (10) become

$$-2a - b - c = 0$$
$$4a - 2b - 3c = 0$$

Then $-4b = 5c$, $8a = c$, so that $-4b = 40a$, $b = -10a$. Then

$$a(x + 1) + b(y - 2) + c(z - 3)$$
$$= a(x + 1) - 10a(y - 2) + 8a(z - 3) = 0$$

and $x - 10y + 8z - 1 + 20 - 24 = x - 10y + 8z - 3 = 0$ is an equation of the plane. *Ans.* $x - 10y + 8z = 3$.

EXERCISE

Find an equation of the plane through P_1, P_2, P_3 for each set of points listed in Exercise 1 of Sec. 2.

4. Parallel planes. Parallel planes have the same normal line. Then every plane parallel to $ax + by + cz + d = 0$ is obtained if we leave a, b, c fixed and vary d.

If we put an equation of a plane in normal form $\lambda x + \mu y + \nu z = p$, we obtain all planes parallel to the given plane and on the same side of the origin by varying $p \geq 0$. The distance between the given plane and a second plane $\lambda x + \mu y + \nu z = p_1$

is then clearly $|p - p_1|$. If, however, we consider all planes parallel to the given plane and on the other side of the origin, we see that their equations are $-(\lambda x + \mu y + \nu z) = p_1$. Then the distance between $\lambda x + \mu y + \nu z = p$ and $-(\lambda x + \mu y + \nu z) = p_1$ is $p + p_1$.

ILLUSTRATIVE EXAMPLES

I. Find an equation of a plane which is parallel to the plane $3x - 2y + 4z = 7$ and which contains the point $P = (-2, 3, 3)$.

Solution

The equation may be taken to have the form $3x - 2y + 4z = k$, where k is to be determined so that P is on the plane. Then $-6 - 6 + 12 = 0 = k$ and the required equation is $3x - 2y + 4z = 0$.

II. Find the distance between the plane $3x - 2y + 4z = 7$ and the plane $3x - 2y + 4z = 12$.

Solution

The normal forms of these two planes are obtained by multiplying the equations by $(9 + 4 + 16)^{-\frac{1}{2}} = (29)^{-\frac{1}{2}}$. The distance is then $(29)^{-\frac{1}{2}}12 - (29)^{-\frac{1}{2}}7 = 5(29)^{-\frac{1}{2}}$.

III. Find the distance between the plane $3x - 2y + 4z = 7$ and the plane $6x - 4y + 8z = -15$.

Solution

The distance is $7(29^{-\frac{1}{2}}) + \tfrac{15}{2}(29)^{-\frac{1}{2}} = \tfrac{29}{2}(29)^{-\frac{1}{2}} = \tfrac{1}{2}\sqrt{29}$.

IV. Find the equations in normal form of the two planes that are five units from the plane $-x + 2y + 2z + 3 = 0$ and parallel to it.

Solution

The normal form of an equation of the given plane is obtained by multiplying the given equation by $-(1 + 4 + 4)^{-\frac{1}{2}} = -\frac{1}{3}$ and thus is

$$\frac{x - 2y - 2z}{3} = 1.$$

The two planes are then 6 units and -4 units from the origin where we interpret a negative distance as equivalent to the replacement of λ, μ, ν by $-\lambda, -\mu, -\nu$. The answers are then $x - 2y - 2z = 18$ for the value 6 and $x - 2y - 2z = -12$ for the value -4.

EXERCISES

1. Find an equation of a plane parallel to the given plane and through the given point in the following cases:

(a) $3x - 2y + z = 1$, $(-1, -2, 1)$

(b) $2x + 2y - z = 0$, $(-1, 1, 1)$

(c) $x + y + 3z = 4$, $(-2, 1, 1)$

(d) $2x - y - 2z = 7$, $(3, -1, 3)$

(e) $3x + 4y - 6z = 183$, $(4, 5, 2)$

(f) $18x - 14y + 17z = 987$, $(1, 1, 0)$

(g) $10x - 7y + 6z = 111,968$, $(2, 3, 1)$

2. Find an equation of a plane parallel to the given plane and having x-intercept 3 in each of the cases of Exercise 1.

3. Find an equation of a plane parallel to the given plane and two units further from the origin in each case of Exercise 1 except (e), (f), and (g).

4. Find the distances between the following pairs of planes:

(a) $2x + 2y - z = 3$, $2x + 2y - z = 18$

(b) $2x - y - 2z = 6$, $-2x + y + 2z = 12$

(c) $3x + 2y + 6z = 14$, $3x + 2y + 6z = 21$

(d) $6x - 2y + 3z = 28$, $12x - 4y + 6z = 35$

(e) $18x - 6y - 9z = 35$, $-6x + 2y + 3z = 21$

5. Find equations of pairs of planes four units from the first of the planes in Exercise 4.

5. Distance from a plane to a point. If an equation $ax + by + cz + d = 0$ of a plane is given and a point $P = (x_1, y_1, z_1)$ is given, we require a formula for the distance δ from the plane to P. We first observe that we may divide the equation of the plane by $\pm \sqrt{a^2 + b^2 + c^2}$ to convert it to the normal form

$$\lambda x + \mu y + \nu z = p.$$

Then the equation of the plane parallel to the given plane and δ units further from the origin is $\lambda x + \mu y + \nu z = p + \delta$. Since P is on this plane, we have

(11) $$\delta = \lambda x_1 + \mu y_1 + \nu z_1 - p.$$

If $\delta > 0$, the point P and the origin O are on opposite sides of the plane, and if $\delta < 0$, the point P and the origin are on the same side of the plane.

If the given plane passes through the origin, the meaning of the sign of δ is not clear, and we shall simply use the formula

$$\delta = |\lambda x_1 + \mu y_1 + \nu z_1|.$$

ILLUSTRATIVE EXAMPLES

I. Find the distance from the plane $2x + 2y - z = 3$ to the point $(2, -1, 5)$.

Solution

The distance is $\frac{1}{3}[2(2) + 2(-1) - (5) - 3] = -2$.

II. Find the distance from $2x + 2y - z = 0$ to $(2, -1, 5)$.

Solution

The distance is $\left|\frac{1}{3}(4 - 2 - 5)\right| = 1$.

EXERCISES

1. Find the distances from the following planes to the corresponding points:

(a) $2x + 2y - z = 3$, $(5, 3, -2)$

(b) $-2x + y + 2z = -2$, $(2, 1, -2)$

(c) $3x + 2y + 6z = 14$, $(1, -2, 1)$

(d) $12x - 4y + 6z = 35$, $(-1, -1, 1)$

(e) $18x - 9y - 9z = 14$, $(1, 1, 3)$

(f) $-6x + 2y + 3z = 0$, $(-1, 2, -3)$

2. Find the z coordinate of a point $P = (-3, 2, z)$ if the distance from the plane $3x + 2y + 6z = 7$ to the point P is 2.

3. Solve Exercise 2 if the distance is -4.

6. Angle between two planes. The angle θ between two planes is defined to be the angle between the rays from O normal to the two planes. It is not uniquely defined if either plane passes through the origin. If the equations are $\lambda x + \mu y + \nu z = p$ and $\lambda_1 x + \mu_1 y + \nu_1 z = p_1$, then $\cos \theta = \lambda\lambda_1 + \mu\mu_1 + \nu\nu_1$. When the equations are given in general form, they must first be converted into normal form.

EXERCISES

1. Compute the cosine of the angle between the following pairs of planes:

(a) $2x + 2y - z = 3$, $3x - 2y + 6z = 14$

(b) $x + y + z = 1$, $2x - y - 2z = 6$

(c) $x - y + z = 2$, $2x + y = 4$

(d) $x + y = 3$, $3x + 4y = 2$

2. Show that the following pairs of planes are perpendicular.

(a) $2x + 2y - z = 3$, $x - 2y - 2z = 1$

(b) $x + y + z = 1$, $x - z = 6$

(c) $3x - 2y + 4z = 1$, $2x + y - z = 1$

7. The line of intersection of two planes. Two nonparallel planes intersect in a line. If the planes have as equations

$$(12) \quad \begin{aligned} a_1x + b_1y + c_1z + d_1 &= 0 \\ a_2x + b_2y + c_2z + d_2 &= 0, \end{aligned}$$

then their line of intersection is orthogonal to both of their normals and the direction cosines of the line satisfy the equations

$$(13) \quad \begin{aligned} a_1\lambda + b_1\mu + c_1\nu &= 0 \\ a_2\lambda + b_2\mu + c_2\nu &= 0. \end{aligned}$$

However, the simplest method of determining the line is to find two points on it. The procedure is illustrated below.

ILLUSTRATIVE EXAMPLE

Give a set of equations in symmetric form of the line L of intersection of the planes

$$2x + y - 2z = 1, \qquad 5x + 4y - 6z = -2$$

and determine direction cosines of L.

Solution

We solve for x and y by writing $8x + 4y - 8z = 4$; therefore $3x - 2z = 6$ and

$$x = \frac{2z + 6}{3}, \qquad y = 2z - 2x + 1 = \frac{2z - 9}{3}.$$

Put $z = 0$, and obtain $P_1 = (2, -3, 0)$ on L.
Put $z = 3$, and obtain $P_2 = (4, -1, 3)$ on L. Then L has as its equation

$$\frac{x - 2}{2} = \frac{y + 3}{2} = \frac{z}{3}$$

and the corresponding direction cosines are

$$\frac{2}{\sqrt{17}}, \qquad \frac{2}{\sqrt{17}}, \qquad \frac{3}{\sqrt{17}}.$$

EXERCISE

Give a set of equations in symmetric form of the line intersection of the following two planes, and determine its direction cosines.

(a) $3x - 2y + z = 5$, $x + y + z = 1$
(b) $x - 2y + z = 6$, $x - z = 2$
(c) $3x + 4y + 2z = 1$, $x - y + z = 6$
(d) $6x + 2y - 3z = -5$, $x - y + z = 6$
(e) $4x - y + z = 2$, $2x + y - 3z = 4$
(f) $x - y - 3z = 2$, $2x + 2y - z = -3$
(g) $4x + 3y - 2z = 1$, $x + 2y - 3z = 4$

8. Angle between a line and a plane. Let the plane $ax + by + cz + d = 0$ and the line

$$(14) \qquad \frac{x - x_1}{a_1} = \frac{y - y_1}{b_1} = \frac{z - z_1}{c_1}$$

be given. We require the angle ϕ between the line and the plane. This angle is defined to be the complement of the angle θ between the directed normal to the plane and the line and is not unique unless the line is also directed. When both are directed, we will have

$$(15) \qquad \frac{x - x_1}{\lambda_1} = \frac{y - y_1}{\mu_1} = \frac{z - z_1}{\nu_1}, \qquad \lambda x + \mu y + \nu z = p,$$

where we have now made λ, μ, ν and λ_1, μ_1, ν_1 unique. Then $\cos \theta = \sin \phi$ and

$$(16) \qquad \sin \phi = \lambda \lambda_1 + \mu \mu_1 + \nu \nu_1.$$

EXERCISE

Find the sine of the angle between the following lines and planes:

(a) $3x - 2y + 4z = 1$, $3x + 4 = 2y - 1 = 6z + 7$

(b) $x + 2y - 2z = 2$, $\dfrac{x}{2} = \dfrac{y}{3} = \dfrac{z}{6}$

(c) $3x + 2y - 6z = 4$, $\dfrac{x - 1}{2} = \dfrac{y}{2} = z - 3$

9. Pencils of planes. Let the equations $f(x, y, c) \equiv a_1 x + b_1 y + c_1 z + d_1 = 0$ and $g(x, y, z) \equiv a_2 x + b_2 y + c_2 z + d_2 = 0$ define two distinct planes. Then the equation

$$(17) \qquad sf(x, y, z) + tg(x, y, z) = 0$$

is called the equation of the *pencil* of planes determined by the two given planes. Indeed, formula (17) is an equation of a plane for all real numbers s and t which are such that the vector $s(a_1, b_1, c_1) + t(a_2, b_2, c_2) \neq 0$. When the two given planes are not parallel, they intersect in a line L; formula (17) represents a plane $Q(s, t)$ for all real numbers s and t not both zero, and every such plane $Q(s, t)$ contains the line L. For a point $P_0 = (x_0, y_0, z_0)$ is on L if and only if both $f_0 = f(x_0, y_0, z_0) = 0$ and $g_0 = g(x_0, y_0, z_0) = 0$; whence $sf_0 + tg_0 = 0$ and P_0 is on $Q(s, t)$.

When the two given planes are parallel, there exists a nonzero real number k such that $(a_2, b_2, c_2) = k(a_1, b_1, c_1)$ but $d_2 \neq kd_1$, since the planes defined by $f(x, y, z) = 0$ and $g(x, y, z) = 0$ have been assumed to be distinct. In this case formula (17) becomes

$$(18) \qquad (s + kt)(a_1x + b_1y + c_1z) + sd_1 + td_2 = 0,$$

and this equation represents a plane for all real numbers s and t such that $s \neq -kt$. All such planes are evidently parallel to the two given planes, and we call the pencil defined by the two given planes a *parallel pencil*.

Conversely, let Q be any plane through the line of intersection L in case the given planes intersect and parallel to $f(x, y, z) = 0$ in the case of a parallel pencil. Then Q is uniquely determined by the property just described and the assumption that Q contains a point $P_0 = (x_0, y_0, z_0)$ *not* on both of the given planes. The corresponding values $f_0 = f(x_0, y_0, z_0)$ and $g_0 = g(x_0, y_0, z_0)$ determine a member $Q(g_0, -f_0)$ of the pencil, and the given plane Q actually coincides with $Q(g_0, -f_0)$. For $sf_0 + tg_0 = 0$, when $s = g_0$ and $t = -f_0$, and P_0 is on $Q(g_0, -f_0)$. It then remains only to see that the given values of s and t actually define a plane. This is true in the case of intersecting planes, since in that case we merely require that s and t should not both be zero, and we have already assumed that f_0 and g_0 are not both zero. In the case of parallel planes, the hypothesis that $s + kt = 0$ implies that $g_0 = kf_0$. Then f_0 cannot be zero, and formula (18) for (x, y, z) replaced by (x_0, y_0, z_0) yields $sd_1 + td_2 = g_0d_1 - f_0d_2 = f_0(kd_1 - d_2) = 0$, $d_2 = kd_1$ contrary to hypothesis. Thus we have shown that the pencil of planes defined by two distinct planes contains all planes through their line of intersection when they intersect, and contains all planes parallel to them when they are parallel.

When two given planes intersect, their equations may be put into normal form and thus written as

$$(19) \quad \lambda_1x + \mu_1y + \nu_1z - p_1 = 0, \qquad \lambda_2x + \mu_2y + \nu_2z - p_2 = 0.$$

Then the pencil of planes defined in this case contains the two planes whose defining equations are

$$(20) \quad (\lambda_1 - \lambda_2)x + (\mu_1 - \mu_2)y + (\nu_1 - \nu_2)z - (p_1 - p_2) = 0,$$
$$(21) \quad (\lambda_1 + \lambda_2)x + (\mu_1 + \mu_2)y + (\nu_1 + \nu_2)z - (p_1 + p_2) = 0.$$

Each of these planes consists of points $P = (x, y, z)$ whose distances $d_1 = \lambda_1x + \mu_1y + \nu_1z - p_1$ and $d_2 = \lambda_2x + \mu_2y + \mu_2z$

$- p_2$ from the two given planes are equal in absolute value. The planes are then the *bisector planes* of the dihedral angles formed by the two given intersecting planes.

ILLUSTRATIVE EXAMPLES

I. Find an equation of the plane containing the line of intersection of the planes $3x - 2y + 4z - 6 = 0$, $2x + 3y + z - 12 = 0$ as well as the point $(-1, 2, 4)$.

Solution

We compute $f(-1, 2, 4) = -3 - 4 + 16 - 6 = 3$, and $g(-1, 2, 4) = -2 + 6 + 4 - 12 = -4$. This yields the values $s = 4$, $t = 3$, and we compute $12x - 8y + 16z - 24 + 6x + 9y + 3z - 36$ to obtain the solution $18x + y + 19z - 60 = 0$.

II. Find an equation of the plane containing the line of intersection of the planes of Example I and normal to a line whose direction numbers (a, b, c) have the property $a = 3b$.

Solution

The coefficients of the required plane are $a = 3 + 2t$, $b = -2 + 3t$, $c = 4 + t$, $d = -6 - 12t$. Then $a = 3 + 2t = 3b = -6 + 9t$, $7t = 9$, $t = \frac{9}{7}$. Multiply by 7 and obtain $21x - 14y + 28z - 42 = 18x + 27y + 9z - 108 = 0$, and the answer is $39x + 13y + 37z - 150 = 0$.

III. Find the bisectors of the dihedral angles formed by the planes $2x + y - 2z = 4$, $2x - 3y + 6z = -2$.

Solution

The required equations are

$$\frac{2x + y - 2z - 4}{3} = \pm \frac{2x - 3y + 6z + 2}{7},$$

that is, $14x + 7y - 14z - 28 = \pm(6x - 9y + 18z + 6)$. Hence, the solutions are $8x + 16y - 32z - 34 = 0$ and $20x - 2y + 4z - 22 = 0$.

EXERCISES

1. Find an equation of the plane containing the line of intersection of the two given planes and the given point in the following cases:

(a) $2x - 2y + z = 3$, $x + 2y + 2z = 5$, $P = (-1, 2, -2)$
(b) $3x + 2y + 6z = 1$, $2x - 6y + 3z = -4$, $P = (2, -1, -3)$
(c) $x - 2y - 2z = 3$, $3x - 2y - 6z = 4$, $P = (-1, 0, 1)$
(d) $3x + 6y + 2z = -3$, $2x + 2y - z = -2$, $P = (1, 1, 1)$
(e) $x + y + z = -4$, $2x - y - z = -3$, $P = (-3, 2, -1)$
(f) $x + z = -2$, $y - 2z = 4$, $P = (1, 2, -3)$

(g) $x + y + z + 1 = 0, 2x - 3y - 3z - 3 = 0, P = (3, -1, -2)$

(h) $x = y + z, y = x + z, P = (4, -1, 2)$

(i) $2x + 4y - 3z = 7, 4x + 2y - 3z = 9, P = (-1, -1, -2)$

(j) $3x - 4y + 6z = 1, 5x - 2y + z = 8, P = (0, 0, 0)$

2. Find an equation $ax + by + cz + d = 0$ of the pencils defined by Exercise 1 in which $2a = 3b$, and find one in which $4d = 3c$.

3. Find the bisectors of the dihedral angles formed by the pairs of planes of parts (a), (b), (c), (d), (e), (f), and (h) of Exercise 1.

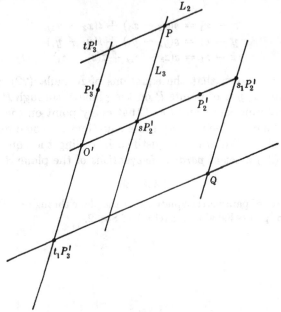

Fig. 10.

10. Parametric equations of a plane. Three points $P_1 = (x_1, y_1, z_1)$, $P_2 = (x_2, y_2, z_2)$, and $P_3 = (x_3, y_3, z_3)$ are not collinear if the vectors

$$P_2' = P_2 - P_1 = (x_2 - x_1, y_2 - y_1, z_2 - z_1), \qquad P_3' = P_3 - P_1$$
$$= (x_3 - x_1, y_3 - y_1, z_3 - z_1)$$

do not have proportional coordinates. Then P_1, P_2, P_3 determine a plane S. Translate the origin to $O' = P_1$, and thus write $P' = (x - x_1, y - y_1, z - z_1)$ for the vector of the coordinates of any points $P = (x, y, z)$. If s and t are any real numbers, the line joining O' to P_3' passes through tP_3'. Hence, sP_2' and tP_3'

are in the plane S and their sum $sP_2' + tP_3'$ is the end of the diagonal of a parallelogram whose vertices are in S; $sP_2' + tP_3'$ is the vector $(x - x_1, y - y_1, z - z_1)$ of a point of S.

Conversely, if P is a point of S, we draw a line L_2 in S through P parallel to P_1P_2 and a line L_3 in S through P parallel to P_1P_3. Then L_3 cuts P_1P_2 in a point whose transformed vector is sP_2' and L_2 cuts P_1P_3 in a point whose transformed vector is tP_3'. For a typical case, see Fig. 10. But then $P' = sP_2' + tP_3'$, and we have

$$(22) \qquad \begin{aligned} x - x_1 &= s(x_2 - x_1) + t(x_3 - x_1) \\ y - y_1 &= s(y_2 - y_1) + t(y_3 - y_1) \\ z - z_1 &= s(z_2 - z_1) + t(z_3 - z_1). \end{aligned}$$

We have shown that the equations of formula (22) give the coordinates x, y, z of points P on the plane S through P_1, P_2, P_3 for all real numbers s and t and that every point on the plane S has coordinates which satisfy these equations for some real numbers s and t. We are then justified in calling the equations of formula (21) a set of *parametric equations* of the plane S.

EXERCISE

Give a set of parametric equations for the plane through P_1, P_2, P_3 for each set of points listed in Exercise 1 of Sec. 2.

CHAPTER 3
SURFACES AND CURVES

1. Equations of a surface. A *real single-valued function* $f(x, y, z)$ of three real variables x, y, z is a correspondence wherein every triple of real numbers x_1, y_1, z_1 uniquely determines a real number $f(x_1, y_1, z_1)$ called the *value* of the function at the point where $x = x_1, y = y_1, z = z_1$. The functions commonly considered in analytic geometry are usually defined by algebraic or trigonometric formulas.

If $f(x, y, z)$ is a real single-valued function, a *point* $P = (x_1, y_1, z_1)$, defined by real coordinates x_1, y_1, z_1, is said to be a *solution* of the equation

$$(1) \qquad\qquad f(x, y, z) = 0$$

if $f(x_1, y_1, z_1)$ is the real number zero. An equation (such as $x^2 + y^2 + z^2 = 0$) may have only a single point as a solution or no real points as solutions. However, the solutions usually make up a geometric locus (*i.e.*, set) of points called a *surface*.

The equation $f(x, y, z) = 0$ is said to be an equation of a surface S if S is the set of all solutions of the equation. A surface, such as a cone, a cylinder, or a sphere, may be defined geometrically, and we shall give geometric definitions of certain special types of surfaces. If we then give an equation $f(x, y, z) = 0$ and state that it is an equation of S, the proof of the result requires that we verify the following two properties:

1. *Every point* P *on* S *is a solution of* f(x, y, z) = 0.
2. *Every solution* P *of* f(x, y, z) = 0 *is a point on* S.

We shall frequently define a surface S by prescribing an equation $f(x, y, z) = 0$ and stating that S is the surface of all points P which are solutions of the equation. As we have seen, the solutions might consist of a single point or no real points and the corresponding surface will then be called a *point surface* or an *imaginary surface*. The general geometric definition of those point sets which are true surfaces is beyond the scope of the present text.

If a is any nonzero real number and $f(x, y, z) = 0$ is an equation of a surface, then $af(x, y, z) = 0$ is also an equation of the same surface. Indeed, let $g(x, y, z)$ be any function such as $x^2 + 1$ which has the property that $g(x_1, y_1, z_1) \neq 0$ for all real numbers x_1, y_1, z_1. Then if $f(x, y, z) = 0$ is an equation of S, so is $g(x, y, z)f(x, y, z) = 0$. For any product $g(x_1, y_1, z_1)f(x_1, y_1, z_1)$ is zero if and only if $f(x_1, y_1, z_1)$ is zero.

It will be convenient for us to speak of "the surface $f(x, y, z) = 0$" when we mean the surface defined by the equation $f(x, y, z) = 0$. Thus we may speak of "the plane $ax + by + cz + d = 0$," and we have already used this terminology in Chap. 1 and a similar terminology for lines in Chap. 2.

2. Space curves. The *intersection* of two surfaces S_1 and S_2 is the locus of all points on both S_1 and S_2. This locus may be a surface; for example, the equation $xy = 0$ is an equation of a surface S which consists of two planes, and the equation $xz = 0$ is the equation of a surface which also consists of two planes. Their intersection is the plane $x = 0$.

In most of the cases commonly considered in analytic geometry the intersection C of two surfaces S_1 and S_2 is a *space curve*. This is a geometric manifold with one degree of freedom and the precise definition is beyond the scope of the present text. We shall assume that the equations

$$f(x, y, z) = 0, \qquad g(x, y, z) = 0$$

are equations of two surfaces and shall call the pair of equations a set of equations of the space curve which is their intersection. The intersection may then turn out to be empty, a single point, a genuine space curve, or a surface, and a revision of the geometric term *space curve*, by means of which the intersection has been described, may be necessary. Note that the curve of intersection of two planes is a straight line and that a straight line is an instance of a space curve.

As in Sec. 1, we shall speak of "the curve $f(x, y, z) = 0$, $g(x, y, z) = 0$" when we wish to refer to the curve of intersection of the surfaces defined by the equations $f(x, y, z) = 0$ and $g(x, y, z) = 0$.

3. Plane sections. The curve of intersection of a plane and a surface is called a *plane section* of the surface. In general, it is an ordinary plane curve, but it may consist of a single point or of no points.

The determination of a plane section may be carried out most simply if an equation of the surface can be obtained relative to a coordinate system in which the given plane is one of the coordinate planes. For example, if the given plane is the x, y plane, and a corresponding equation of the surface is $f(x, y, z) = 0$, then the plane section is the curve $f(x, y, 0) = 0$ in the x, y plane.

When a surface is given by means of an equation $f(x, y, z) = 0$ and a plane by an equation $ax + by + cz + d = 0$, it is always possible to perform a transformation of coordinates so as to obtain an x', y', z' coordinate system in which the given plane becomes the plane $z' = 0$ and the surface is given relative to the transformed coordinate system by an equation $\phi(x', y', z') = 0$. The plane section may then be analyzed by a study of the equation $\phi(x', y', 0) = 0$. The equations that transform $f(x, y, z)$ into $\phi(x', y', z')$ will be developed in Chap. 7.

ILLUSTRATIVE EXAMPLE

Discuss the sections of the surface

$$\frac{x^2}{16} + \frac{y^2}{4} + \frac{z^2}{9} = 1$$

made by planes parallel to the x, y plane.

Solution

The required planes are the planes $z = k$, where k is a real number. Then the sections are curves whose equations are

$$\frac{x^2}{16} + \frac{y^2}{4} = \frac{9 - k^2}{9}$$

If $k^2 > 9$, the corresponding plane does not intersect the surface. When $k = 3$, the intersection is the single point $(0, 0, 3)$ and the plane is said to be tangent to the surface. Similarly, if $k = -3$, the plane $z = -3$ is tangent to the surface at the point $(0, 0, -3)$. If $k^2 < 9$, the corresponding sections are ellipses having center at $(0, 0, k)$, semimajor axes $\frac{4}{3} \sqrt{9 - k^2}$, semiminor axes $\frac{2}{3} \sqrt{9 - k^2}$, and a major axis, which is a segment of the line $z = k$, $y = 0$, parallel to the x axis.

EXERCISE

Discuss the sections of the following surfaces made by planes parallel to the x, y plane:

(a) $\dfrac{x^2}{4} + \dfrac{y^2}{4} + \dfrac{z^2}{9} = 1$ (e) $2x^2 - y^2 - z^2 = 1$

(b) $\dfrac{x^2}{9} + \dfrac{y^2}{4} - \dfrac{z^2}{16} = 0$ (f) $4x^2 + 2y^2 + 4z^2 = 1$

(c) $x^2 = 4yz$ (g) $4zx = y^2 + 2y$

(d) $\dfrac{x^2}{16} + \dfrac{y^2}{9} - \dfrac{z^2}{4} = 1$ (h) $x^2 + 2x + 4y^2 - z^2 + 2z = 0$

4. Algebraic surfaces. A surface is called an *algebraic surface* if it is defined by an equation $f(x, y, z) = 0$, where $f(x, y, z)$ is a polynomial in x, y, z with real coefficients. The curve of intersection of two algebraic surfaces is called an *algebraic curve.*

Every polynomial $f(x, y, z)$ is a sum of terms of the form

$$ax^r y^s z^t,$$

where the coefficient a is a real number and r, s, t are natural numbers, *i.e.*, nonnegative integers. The degree of such a term is $r + s + t$, and the degree n of $f(x, y, z)$ is the largest degree of all of the terms of $f(x, y, z)$ that have nonzero coefficients. We call $f(x, y, z)$ a constant polynomial if $n = 0$. If $n = 1$, the polynomial $f(x, y, z) = ax + by + cz + d$ and the equation $f(x, y, z) = 0$ is an equation of a surface that we have called a *plane*. If $n = 2$, we call $f(x, y, z) = 0$ a *quadric* surface, and if $n = 3$ we call $f(x, y, z) = 0$ a *cubic* surface.

If $f(x, y, z) = g(x, y, z)h(x, y, z)$, where $g(x, y, z)$ and $h(x, y, z)$ are polynomials with real coefficients and neither is a constant, we call $f(x, y, z)$ a *reducible* polynomial. Otherwise, we call $f(x, y, z)$ an *irreducible* polynomial. In the former case $f(x, y, z) = 0$ is an equation of the surface consisting of all solutions of either of the equations.

$$g(x, y, z) = 0, \qquad h(x, y, z) = 0.$$

If either of these surfaces is an imaginary locus, such as $x^2 + y^2 + z^2 + 1 = 0$, the factor is deleted and $f(x, y, z)$ is replaced by the remaining factor. However, when both surfaces contain real points, we call the surface defined by $f(x, y, z) = 0$ a reducible surface. Then every irreducible surface is defined by an equation $f(x, y, z) = 0$, where $f(x, y, z)$ is an irreducible polynomial.

5. Cones. A *cone* is a surface S containing a fixed point V called the *vertex* of S such that all points of any line joining V to a point of S are points of S. We shall restrict our attention to algebraic cones, *i.e.*, cones which are algebraic surfaces and shall assume that $n > 1$. Thus we do not include planes in the set of surfaces called *cones*.

A polynomial $f(x, y, z)$ is said to be *homogeneous* if all terms of $f(x, y, z)$ have the same degree n. A homogeneous polynomial of degree n is called an *n*ic form. We shall study quadratic forms, *i.e.*, forms of degree two, later. If $f(x, y, z)$ is homogeneous of degree n, then $f(tx, ty, tz) = t^n f(x, y, z)$. Also $f(0, 0, 0) = 0$ if $n > 0$.

Theorem 1. *If* $f(x, y, z)$ *has the property that* $g(x', y', z') \equiv f(x' + x_0, y' + y_0, z' + z_0)$ *is homogeneous of degree* $n > 1$ *in* x', y', z' *then* $f(x, y, z) = 0$ *is an equation of a cone with vertex at* (x_0, y_0, z_0).

For a translation of axes, which moves the origin to $V = (x_0, y_0, z_0)$, replaces $f(x, y, z)$ by $g(x', y', z')$. Since $g(0, 0, 0) = 0$, the point V is on the surface S defined by $f(x, y, z) = 0$. If $Q = (x_1', y_1', z_1')$ is a point on S, then $g(x_1', y_1', z_1') = 0$ and $g(tx_1', ty_1', tz_1') = t^n g(x_1', y_1', z_1') = 0$ for every t. But $t(x_1', y_1', z_1') = (x', y', z')$ is the vector form of a set of parametric equations of the line joining V to Q, and we have proved that all points of this line are on S, S is a cone.

The cones, which consist of all points on the lines joining the vertex to the points of a fixed plane curve, are of particular interest in geometry. Let us then derive the equations of such surfaces in the case where the given curve is a curve in the x, y plane and its equation is $\phi(x, y) = 0$. Suppose that $P_0 = (x_0, y_0, z_0)$ is the vertex. Then a point $P = (x, y, z)$ is on the cone if and only if $P - P_0 = t(P_1 - P_0)$, where $P_1 = (x_1, y_1, 0)$ is a point on the curve. Consequently

$$(2) \quad x - x_0 = t(x_1 - x_0), \qquad y - y_0 = t(y_1 - y_0),$$
$$z - z_0 = -tz_0.$$

We assume that P_0 is not a point of the x, y plane and thus $z_0 \neq 0$. Then we may eliminate t and obtain

$$z_0(x - x_0) = (x_1 - x_0)(z_0 - z), \qquad z_0(y - y_0) = (y_1 - y_0)(z_0 - z).$$

We solve for x_1 and y_1 to obtain

$$(3) \qquad x_1 = x_0 + \frac{z_0(x - x_0)}{z_0 - z}, \qquad y_1 = y_0 + \frac{z_0(y - y_0)}{z_0 - z},$$

and substitute these formulas in $\phi(x_1, y_1) = 0$. The resulting equation in x, y, z is an equation of the given cone.

ILLUSTRATIVE EXAMPLES

I. Find an equation of the elliptic cone with vertex at $(2, -1, 3)$, and base the curve $4x^2 + y^2 = 1$.

Solution

We use formula (3) and write

$$x_1 = 2 + \frac{3(x - 2)}{3 - z}, \qquad y_1 = -1 + \frac{3(y + 1)}{3 - z}.$$

Then $(3 - z)x_1 = 3x - 2z$, $(3 - z)y_1 = 3y + z$, so that the required equation is

$$4(3x - 2z)^2 + (3y + z)^2 = (3 - z)^2.$$

The solution should be left in this form since the equation is not essentially simplified when the indicated operations are carried out.

II. Find an equation of the hyperbolic cone through the point $(3, 0, 0)$ and the hyperbola whose equations are $z = 0$, $(x^2/9) - (y^2/4) = 1$.

Solution

We use the method of derivation of formula (3) to write $x - 3 = -3t$, $y = ty_1$, $z = tz_1$. Then $-3y = (x - 3)y_1$, $-3z = (x - 3)z_1$, and

$$\frac{(x - 3)^2 z_1{}^2}{9} - \frac{(x - 3)^2 y_1{}^2}{4} = (x - 3)^2 = \frac{9z^2}{9} - \frac{9y^2}{4}.$$

Hence $4(x - 3)^2 = 4z^2 - 9y^2$ is an equation of the cone.

EXERCISES

1. Find an equation of the cone with vertex at the given point P and passing through the curve whose equations are $z = 0$, $\phi(x, y) = 0$ in the following cases:

(a) $(0, 0, 3)$, $x^2 + y^2 = 9$ (c) $(1, -1, 2)$, $x^2 + y^2 = 1$
(b) $(0, 0, -2)$, $x^2 + y^2 = 4$ (d) $(-1, 1, -3)$, $x^2 - y^2 = 1$

(e) $(-1, 2, -1)$, $3x^2 + 2y^2 = 1$
(f) $(-1, 2, 3)$, $x^2 + 2x + y^2 - 4y + 3 = 0$
(g) $(2, 1, -4)$, $x^2 - 4x + 3y^2 - 6y + 6 = 0$

(h) $(-1, 2, 3)$, $x^2 = 8y$ (j) $(2, 1, 4)$, $x^2 - 3y^2 = 1$
(i) $(0, -1, 2)$, $2x^2 - y^2 = 1$ (k) $(0, 0, 1)$, $x^3 + y^3 = 1$

2. Interchange x and z in the equations of the plane curves of Exercise 1, and leave the points P unaltered. Find equations of the corresponding cones. The equations are then $x = 0$, $\phi(z, y) = 0$.

3. Find an equation of the cone with vertex at $(-1, 1, 2)$ and passing through the curve of intersection of the surface $x^2 + 4y^2 = 1$ and the plane $z = 3$.

6. Cylinders. A *cylinder* is a surface consisting all of the points on all the lines which are parallel to a given line and which pass through a fixed plane curve in a plane not parallel to the given line. It is a simple matter to obtain an equation of a cylinder with respect to a coordinate system which is chosen so that the plane curve lies in a coordinate plane.

Let us assume that the curve lies in the x, y plane so that its equations are $z = 0$, $\phi(x, y) = 0$. The fixed line will have direction numbers a, b, c, and a point $P = (x, y, z)$ will be on the cylinder S if and only if

(4)
$$\frac{x - x_1}{a} = \frac{y - y_1}{b} = \frac{z}{c},$$

where $(x_1, y_1, 0)$ is on the given curve. Then

(5)
$$x_1 = x - \frac{a}{c}z, \qquad y_1 = y - \frac{b}{c}z,$$

and

(6)
$$\phi\left(x - \frac{a}{c}z, \quad y - \frac{b}{c}z\right) = 0$$

is an equation of the surface. Note that our hypothesis that the given line is not parallel to the given plane implies that the number c is not zero.

If $a = b = 0$, the given line is parallel to the z axis and the equation we have derived reduces to $\phi(x, y) = 0$. Thus an equation in two of the three variables x, y, z represents a cylinder generated by a line parallel to the coordinate axis corresponding to the missing variable.

ILLUSTRATIVE EXAMPLES

I. Find an equation of the elliptic cylinder determined by the curve $z = 0$, $4x^2 + y^2 = 1$ and a line with direction numbers $(2, -1, 3)$.

Solution

We use formula (6) to write the answer

$$4(x - \tfrac{2}{3}z)^2 + (y + \tfrac{1}{3}z)^2 = 1.$$

II. Find an equation of the hyperbolic cylinder determined by the curve $x = 5$, $9z^2 - 4y^2 = 1$ and the line with direction numbers $(1, -2, 3)$.

Solution

As in formula (4), the points on the cylinder satisfy

$$\frac{x - 5}{1} = \frac{y - y_1}{-2} = \frac{z - z_1}{3},$$

where $(5, y_1, z_1)$ is on the curve. Then

$$y_1 = y + 2(x - 5), \qquad z_1 = z - 3(x - 5)$$

and thus the required equation is

$$9(z - 3x + 15)^2 - 4(y + 2x - 10)^2 = 1.$$

III. Find an equation of the cylinder determined by the curve of intersection of the plane $2x + 3y - z = 1$ with the surface $3x^2 - y^2 + 2z^2 = 1$ and the line with direction numbers $(-1, 2, 3)$.

Solution

The points $P = (x, y, z)$ satisfy

$$\frac{x - x_1}{-1} = \frac{y - y_1}{2} = \frac{z - z_1}{3},$$

where x_1, y_1, z_1 are on the curve. Then $z_1 = z - 3(x_1 - x)$, $y_1 = y - 2(x_1 - x)$ and $2x_1 + 3y_1 - z_1 = 2x_1 + 3y - 6(x_1 - x) - z + 3(x_1 - x) = 1$. Hence, $x_1 = 3x + 3y - z - 1$, $z_1 = z + 3x - 3(3x + 3y - z - 1) = -6x - 9y + 4z + 3$, $y_1 = y + 2x - 2(3x + 3y - z - 1) = -4x - 5y + 2z + 2$. Then our solution is $3(3x + 3y - z - 1)^2 - (4x + 5y - 2z - 2)^2 + 2(6x + 9y - 4z - 3)^2 = 1$.

EXERCISES

1. Find an equation of a cylinder determined by the given plane curve and set of direction numbers in each of the following cases:

(a) $x^2 + y^2 = 9$, $z = 0$, $(-1, 2, 3)$
(b) $x^2 - y^2 = 4$, $z = 2$, $(2, -1, 4)$
(c) $2y - z^2 = 1$, $x = -3$, $(-1, -1, 1)$
(d) $3x^2 + 2z^2 = 1$, $y = 3$, $(1, 2, 0)$
(e) $x^2 - 8y$, $z = 0$, $(0, 1, 2)$
(f) $y^2 = 4z - 2$, $x = 0$, $(-1, 3, 2)$
(g) $z^2 = 4x^2 + 8x$, $y = 0$, $(0, 4, 1)$
(h) $x^3 + y^3 = 1$, $z = 0$, $(-1, 1, 1)$
(i) $x^4 - y^4 = 1$, $z = 0$, $(-2, 1, -1)$

　(j) $z^3 = x^2$, $y = 3$, $(-2, 2, 1)$
　(k) $y^2 - z^2 = -1$, $x = 2$, $(-2, 2, 2)$
　(l) $x + z^2 = 1$, $y = 0$, $(1, 1, 1)$
　(m) $x^3 - y^3 = xy$, $z = 0$, $(1, 0, 1)$
　(n) $zy = -2$, $x = 1$, $(-2, 1, 1)$

2. In the following cases, the plane curves are not in planes parallel to coordinate planes. Find an equation of each of the corresponding cylinders.

　(a) $x^2 + y^2 = 1$, $x = z$, $(-1, 1, 2)$
　(b) $x^2 + y^2 + z^2 = 1$, $x = 2y$, $(1, 2, 3)$
　(c) $3x^2 - 4y^2 + z^2 = 1$, $x - y = z$, $(1, -1, 1)$
　(d) $x^2 = y + z$, $x = z - 2y$, $(-2, 1, 1)$

7. Surfaces of revolution. If a plane curve (or a line) is revolved about a line in the plane of the curve, the resulting surface is called a *surface of revolution*. We shall obtain the equations of such surfaces relative to a coordinate system where the given plane is the x, y plane and the given line is parallel to the x axis.

Let $\phi(x, y) = 0$, $z = 0$ be equations of the curve and $z = 0$, $y = k$ be equations of the line. Then a point $P = (x, y, z)$ is on the surface of revolution S if and only if P is on a circle whose center is $(x, k, 0)$ and whose radius is the distance $|y_1 - k|$ from $(x, k, 0)$ to the point $(x, y_1, 0)$ on the given curve. Thus $P = (x, y, z)$ is at the same distance from $(x, k, 0)$, and

$$|y_1 - k| = \sqrt{(x - x)^2 + (y - k)^2 + z^2} = \sqrt{(y - k)^2 + z^2}.$$

It follows that $(y_1 - k)^2 = (y - k)^2 + z^2$ and that the required equation of the surface S is obtained by replacing $(y_1 - k)^2$ by $(y - k)^2 + z^2$ in $\phi(x, y) = 0$ or in a suitably modified equation. For the modifications, see the examples below.

ILLUSTRATIVE EXAMPLES

I. Find an equation of the surface of revolution obtained by revolving the curve $z = 0$, $x^2 = 4y$ about the x axis.

Solution

If $(x, y, 0)$ is on the curve both $(x, y, 0)$ and $(x, -y, 0)$ are on the surface, the surface is the surface of revolution of the double curve $x^4 = 16y^2$. Hence, an equation of the required surface is $x^4 = 16(y^2 + z^2)$.

II. Find an equation of the surface of revolution obtained by revolving the curve $z = 0$, $x^2 = 2y$ about the line $z = 0$, $y = 3$.

Solution

The equations of curve are $z = 0$, $x^2 - 6 = 2(y - 3)$, and therefore $(x^2 - 6)^2 = 4(y - 3)^2$ may be converted into the required equation $(x^2 - 6)^2 = 4[(y - 3)^2 + z^2]$, that is, $x^4 - 12x^2 + 36 = 4y^2 - 24y + 4z^2 + 36$. Hence, the answer required is $x^4 - 12x^2 = 4y^2 - 24y + 4z^2$.

EXERCISES

1. Find the equations of the surfaces of revolution obtained by revolving the following curves in the x, y plane about the x axis:

(a) $x^2 + 2y^2 = 1$ (f) $x^2 - 4x + 2y = 1$

(b) $x^2 - y^2 = 4$ (g) $x^4 + 2x^2y + y^2 = 1$

(c) $2x^2 - 3y^2 = 2$ (h) $x^2 + 3y^2 = 2xy$

(d) $x^2 + 2x + y^2 = 1$ (i) $x^3 + y^3 = 1$

(e) $x^2 + xy + y^2 = 1$ (j) $x^2 - y^2 = x^3y^3$

2. Solve Exercise 1 for revolution about the y axis.

3. Solve Exercise 1 for revolution about the following lines in the x, y plane:

(a) $x = -1$ (b) $y = 2$ (c) $x = 1$

8. Symmetries of surfaces. Let S be a surface and T be a plane. Then S is said to be *symmetric with respect to T* if every line perpendicular to T and cutting S in a point P not on T also cuts S in a point Q such that P and Q are on opposite sides of T and the same distance from T. If T is taken to be the x, y plane and $f(x, y, z) = 0$ is the equation of S, then S will be symmetric with respect to T if and only if $f(x, y, -z) = 0$ for every point $P = (x, y, z)$ such that $f(x, y, z) = 0$. This is evidently satisfied when $f(x, y, z)$ is a polynomial in x, y, and z^2, rather than z.

A surface S is said to be *symmetric with respect to a line L* if every line perpendicular to L which cuts S in a point P not on L also cuts S in a second point Q not on L such that P and Q are on opposite sides of L and the same distance from L. If L is taken to be the x axis and $f(x, y, z) = 0$ is an equation of S, then S will by symmetric with respect to L if and only if $f(x, -y, -z) = 0$ for every point $P = (x, y, z)$ such that $f(x, y, z) = 0$. This is satisfied when $f(x, y, z)$ is a polynomial in x, y^2, z^2, yz.

A surface S is said to be *symmetric with respect to a point O* if every line joining O to a point P on S also joins O to a point Q on S such that $\overrightarrow{QO} = \overrightarrow{OP}$. If O is the origin, $Q = -P$ and therefore we must have $f(-x, -y, -z) = 0$ for every point $P = (x, y, z)$ such that $f(x, y, z) = 0$. In the exercises below only symmetries with respect to planes parallel to coordinate planes and lines parallel to coordinate axes are to be discussed.

ILLUSTRATIVE EXAMPLES

I. Discuss the symmetries of the surface $x^2 - 4x + 3y^2 + 2z^2 - 12z = 1$.

Solution

We translate the origin so as to simplify the equation. The equation is $(x - 2)^2 + 3y^2 + 2(z - 3)^2 = 23$, and thus the translation $x' = x - 2, y' = y, z' = z - 3$ converts the equation to $x'^2 + 3y'^2 + 2z'^2 = 23$. This surface has symmetry with respect to the new origin, the $x', y',$ and z' planes, and the $x', y',$ and z' axes. Thus the original surface is symmetric with the planes $x = 2, y = 0,$ and $z = 3$, the lines of intersection of these planes, and their point of intersection $(2, 0, 3)$.

II. Discuss the symmetries of the surface $2x^2 + 4x + 3yz^3 + 1 = 0$.

Solution

The equation is equivalent to $2(x + 1)^2 + 3yz^3 = 1$, and thus to $2x'^2 + 3y'z'^3 = 1$, where $x' = x + 1, y' = y, z' = z$. This equation is unaltered when we replace x' by $-x'$ and y', z' by $-y', -z'$. It is then a surface symmetric with respect to the point $(-1, 0, 0)$, the plane $x = -1$, and the line $y = z = 0$.

EXERCISES

Discuss the symmetries of the following surfaces:

(a) $3x^2 + y^2 - z = 0$

(b) $x^2 - 2yz = 1$

(c) $x^3 + y^3 = z^3$

(d) $x^3 = y^3z^4$

(e) $x^2 - 2x = yz$

(f) $3x^3 - x = yz$

(g) $x^4 + 2x^2 = y^2z$

(h) $x^2 - x^3 = y^2 + yz + z^2$

(i) $x^2 = y^2 - 2y + z(y - 1)$

(j) $x^3 = (y^2 - 2y)z + z^3$

(k) $x^3 + y^4 - z^5 = 0$

(l) $x^2 + y^2z^2 = x^3y$

9. Intersections of a line and a surface. Every line L is defined by a set of equations

(7) $x = x_0 + t\xi,$ $y = y_0 + t\eta,$ $z = z_0 + t\zeta,$

where (x_0, y_0, z_0) is a point on L, (ξ, η, ζ) is a set of direction numbers of L, and any real value of t defines a corresponding point (x, y, z) of L. If $f(x, y, z) = 0$ is an equation of a surface S, the points of intersection of L and S are those points of L for which

(8) $\phi(t) = f(x_0 + t\xi,\quad y_0 + t\eta,\quad z_0 + t\zeta) = 0.$

If $f(x, y, z) = 0$ is the equation defining an algebraic surface of degree n, the corresponding function $\phi(t)$ is a polynomial in t whose degree cannot exceed n. Then either $\phi(t)$ is identically zero and all values of t satisfy $\phi(t) = 0$ or $\phi(t) = 0$ has at most n real roots. In the former case, all points of L are points of S, and the line L lies wholly on the surface S. We have proved the following result:

Theorem 2. *If a line is not wholly on an algebraic surface of degree* n, *it cuts the surface in at most* n *points.*

A line then cuts a quadric surface in no points, one point, or two points or is a line all of whose points are on the surface.

ILLUSTRATIVE EXAMPLE

Find the points of intersection of the line joining $(-1, 2, 3)$, $(2, -1, 4)$, and the surface $x^2 - y^2 + 2z^2 + 1 = 0$.

Solution

We write $x = -1 + 3t$, $y = 2 - 3t$, $z = 3 + t$ and have $(-1 + 3t)^2 - (2 - 3t)^2 + 2(3 + t)^2 + 1 = 9t^2 - 6t + 1 - (9t^2 - 12t + 4) + 2(t^2 + 6t + 9) + 1 = 2t^2 + 18t + 16 = 0$, $t^2 + 9t + 8 = (t + 8)(t + 1) = 0$. Hence, $t = -8, -1$, and the points are $(-25, 26, -5)$, $(-4, 5, 2)$.

EXERCISE

Find the points of intersection of the lines joining P_1 to P_2 and the surface $f \equiv f(x, y, z) = 0$ in the following cases:

(a) $P_1 = (0, 0, 0)$, $P_2 = (2, -1, 3)$, $f \equiv 2x^2 + y^2 - z^2 + 1$

(b) $P_1 = (0, 0, 0)$, $P_2 = (2, -1, 3)$, $f \equiv 2x^2 + y^2 - z^2$

(c) $P_1 = (1, 1, 2)$, $P_2 = (3, 0, 5)$, $f \equiv 2(x - 1)^2 + 2(y - 1)^2 - (z - 2)^2 - 1$

(d) $P_1 = (-1, 2, 3)$, $P_2 = (1, 1, 2)$, $f = x^2 + y^2 - 5z^2 + 21$

(e) $P_1 = (-4, 2, 3)$, $P_2 = (-3, 3, 4)$, $f \equiv x^2 + y^2 + z^2 - 2$

(f) $P_1 = (-1, 2, 3)$, $P_2 = (1, 1, 2)$, $f \equiv 3x^2 - 8y^2 - 4z^2 + 41$

(g) $P_1 = (-1, 1, 2)$, $P_2 = (3, -4, 1)$, $f \equiv x^2 + 2y^2 + 3z^2 - 7$

10. Lines on a cylinder. The lines on a cylinder, consisting of lines parallel to the z axis and passing through a curve in the x, y plane, may be determined easily. We first prove the following:

Lemma 1. *Let* f(x, y) *be a homogeneous polynomial and* α, β *be real numbers not both zero. Then* $f(\alpha, \beta) = 0$ *if and only if* $\beta x - \alpha y$ *is a factor of* f(x, y).

For if $f(x, y) \equiv (\beta x - \alpha y)g(x, y)$, then $f(\alpha, \beta) = (\beta \alpha - \alpha \beta)$. $g(\alpha, \beta) = 0$. Conversely, let $f(\alpha, \beta) = 0$, and define $\xi = \beta x - \alpha y$, $\eta = \alpha x + \beta y$. Since $\alpha^2 + \beta^2 \neq 0$, we see by direct computation that

$$x = \frac{\beta \xi + \alpha \eta}{\alpha^2 + \beta^2}, \qquad y = \frac{-\alpha \xi + \beta \eta}{\alpha^2 + \beta^2}.$$

Then $f(x, y) \equiv \phi(\xi, \eta)$, where $\phi(\xi, \eta) = a_0 \xi^n + a_1 \xi^{n-1} \eta + \cdots + a_n \eta^n$ is necessarily homogeneous in ξ and η. When $x = \alpha$ and $y = \beta$, we have $\xi = \beta \alpha - \alpha \beta = 0$, $\eta = \alpha^2 + \beta^2$, and $f(\alpha, \beta) = \phi(0, \alpha^2 + \beta^2) = a_n(\alpha^2 + \beta^2)^n = 0$. Then $a_n = 0$, $\phi(\xi, \eta) \equiv \xi \cdot \psi(\xi, \eta) \equiv f(x, y) \equiv (\beta x - \alpha y)g(x, y)$, where $g(x, y) \equiv \psi(\beta x - \alpha y, \alpha x + \beta y)$. This proves the lemma.

We use the lemma in the proof of the following theorem:

Theorem 3. *Let* S *be a cylinder defined by an irreducible polynomial equation* f(x, y) = 0, *of degree* n > 1. *Then there is only one line of* S *through each point of* S.

For if P is any point of S, we may translate the origin to P and therefore assume that $f(0, 0) = 0$. Then $f(x, y) = f_n(x, y) + f_{n-1}(x, y) + \cdots + f_1(x, y)$ where each $f_k(x, y)$ is homogeneous of degree k. Every line L through the origin P has equations $x = t\alpha, y = t\beta, z = t\gamma$, and

$$\phi(t) = f(t\alpha, t\beta) = t^n f_n(\alpha, \beta) + t^{n-1} f_{n-1}(\alpha, \beta) + \cdots + t f_1(\alpha, \beta).$$

Then L is on S if and only if $\phi(t) \equiv 0$, that is, $f_n(\alpha, \beta) = f_{n-1}(\alpha, \beta) = \cdots = f_1(\alpha, \beta) = 0$. If $(\alpha, \beta) \neq (0, 0)$, Lemma 1 implies that $\beta x - \alpha y$ is a factor of every $f_k(x, y)$ and thus is a factor of $f(x, y)$ contrary to our hypothesis that $n > 1$ and that $f(x, y)$ is irreducible. Hence, $\alpha = \beta = 0$ and the only line through P is the z axis.

EXERCISE

Show that a surface $f(x, y) = 0$ determined by a homogeneous polynomial $f(x, y)$ of degree n consists of $r \leq n$ planes through the z axis or is a point surface.

11. Tangent lines and planes. A line $x = x_0 + t\xi$, $y = y_0 + t\eta$, $z = z_0 + t\zeta$ will intersect a surface $f(x, y, z) = 0$ in a point $P_1 = (x_1, y_1, z_1)$ if and only if there exists a real number t_1 which is a root of the equation $\phi(t) = f(x_0 + t\xi, y_0 + t\eta, z_0 + t\zeta) = 0$ such that $x_1 = x_0 + t_1\xi$, $y_1 = y_0 + t_1\eta$, $z_1 = z_0 + t_1\zeta$. We shall say that the line is tangent to the surface at the point (of tangency) P_1 if t_1 is a root of multiplicity $m \geq 2$ of $\phi(t) = 0$.

Let us investigate the problem of determining whether or not a given line through a point (x_0, y_0, z_0) on a surface $f(x, y, z) = 0$ is tangent to the surface at the point. The point corresponds to the value $t = 0$ and so we are assuming that $\phi(0) = 0$. We limit our discussion to functions $f(x, y, z)$ having partial derivatives of all orders and see that $\phi(t)$ has a Taylor series expansion

$$\phi(t) = \phi(0) + t\phi'(0) + \frac{t^2}{2} \phi''(0) + \cdots$$

which is an ordinary polynomial when $f(x, y, z)$ is a polynomial. Then the given line is a tangent line if and only if $\phi'(0) = 0$.

The function $\phi'(t)$ may be computed most easily by partial differentiation. Define $f_x(x, y, z)$ to be the partial derivative of $f(x, y, z)$ with respect to x, that is, the derivative of $f(x, y, z)$ as a function of x alone, and define $f_y(x, y, z), f_z(x, y, z)$ similarly. Then the derivative $\phi'(t)$ of $\phi(t)$ with respect to t is shown in elementary calculus to be given by the formula

$$(9) \qquad \phi'(t) = f_x \frac{dx}{dt} + f_y \frac{dy}{dt} + f_z \frac{dz}{dt} = f_x\xi + f_y\eta + f_z\zeta.$$

Since $x = x_0$, $y = y_0$, $z = z_0$ at $t = 0$, and $(\xi, \eta, \zeta) = (x_2, y_2, z_2) - (x_0, y_0, z_0)$, we see that a line joining the point (x_0, y_0, z_0) to a point $P_2 = (x_2, y_2, z_2)$ is tangent to the surface at (x_0, y_0, z_0) if and only if P_2 is a solution of the equation

$$(10) \quad f_x(x_0, y_0, z_0)(x - x_0) + f_y(x_0, y_0, z_0)(y - y_0) \\ + f_z(x_0, y_0, z_0)(z - z_0) = 0.$$

If the coefficients f_x, f_y, f_z are all zero at the point $P_0 = (x_0, y_0, z_0)$, we call P_0 a *singular point* of the surface. All other points are called *ordinary points*. Since formula (10) is an equation of a plane at an ordinary point P_0 we have proved the following result:

Theorem 3. *A line passing through an ordinary point* (x_0, y_0, z_0) *of a surface* $f(x, y, z) = 0$ *is tangent to the surface if and only if it is a line in the plane of formula* (10). *We call this plane the tangent plane to the surface at the given point.*

If P_0 is an ordinary point of two surfaces S_1 and S_2, we define the angle between S_1 and S_2 at P_0 to be the angle θ between the two corresponding tangent planes at P_0. Then S_1 and S_2 are said to be orthogonal at P_0 if $\theta = 90°$, *i.e.*, the tangent planes are perpendicular.

The line perpendicular to the tangent plane of a surface S at an ordinary point P_0 of S is called the *normal* to S at P_0. Equations of this line are

$$(11) \qquad \frac{x - x_0}{f_x(x_0, y_0, z_0)} = \frac{y - y_0}{f_y(x_0, y_0, z_0)} = \frac{z - z_0}{f_z(x_0, y_0, z_0)}.$$

ILLUSTRATIVE EXAMPLES

I. Find an equation of the plane tangent to the surface $3x^3 - 2y^2 + 4z + 1 = 0$ at the point $(-1, 1, 1)$.

Solution

We compute $f_x = 9x_0^2 = 9, f_y = -4y_0 = -4, f_z = 4$ and the required equation is $9(x + 1) - 4(y - 1) + 4(z - 1) = 9x - 4y + 4z + 9 + 4 - 4 = 0.$　　　　*Ans.*　$9x + 4y - 4z + 9 = 0.$

II. Find equations of the normal line to the surface $3x^3 - 2y^2 + 4z + 1 = 0$ at the point $(-1, 1, 1)$.

Solution

By our theory and the result of Example I the answer is

$$\frac{x + 1}{9} = \frac{y - 1}{-4} = \frac{z - 1}{4}.$$

III. The point $(-1, 1, 1)$ is a point of intersection of the surface $3x^3 - 2y^2 + 4z + 1 = 0$ and the surface $2x^2 - 4xy + z^3 = 7$. Find the cosine of the angle θ between the two surfaces at this point.

Solution

The plane tangent to the second surface is $(4x_0 - 4y_0)(x - x_0) - 4x_0(y - y_0) + 3z_0^2(z - z_0) = 0$ and therefore a set of direction numbers of the normal line is $(-8, 4, 3)$. A set of direction numbers of the first line is $(9, -4, 4)$ and therefore

$$\cos \theta = \frac{|-72 - 16 + 12|}{\sqrt{(-8)^2 + (-4)^2 + 3^2} \sqrt{9^2 + (-4)^2 + (4)^2}} = \frac{76}{\sqrt{(89)(113)}}.$$

EXERCISES

1. A homogeneous algebraic equation $f(x, y, z) = 0$ of degree $n \geqq 2$ defines a cone with vertex at the origin. Show that the vertex is a singular point of the cone.

2. Find an equation of the tangent plane to each of the following surfaces at the given points:

(a) $x^3 - 2y^2 + 5z^2 = 18$, $(0, -1, 2)$
(b) $3x^2 + 4y^3 + 2z = 0$, $(0, -1, 2)$
(c) $x^3 - x^2y + z^2 = 3$, $(1, -1, -1)$
(d) $2xy - z^3 = -1$, $(1, -1, -1)$
(e) $xy - 3z = 3$, $(2, 3, 1)$
(f) $xyz = 6$, $(2, 3, 1)$
(g) $x^2z + 4y^2 = 2$, $(-1, 0, 2)$
(h) $xy^2 + yz^2 + zx^2 = 2$, $(-1, 0, 2)$

3. The problems in Exercise 2 may be grouped in pairs such that the given point P_0 is a point on a pair of planes. Find the cosine of the angle between the planes for each pair.

12. Tangents to quadrics. A *quadric surface* is defined by a polynomial equation

$$(12) \quad f(x, y, z) = ax^2 + by^2 + cz^2 + 2(dxy + exz + gyz) \\ + 2(hx + py + qz) + s = 0,$$

where the coefficients $a, b, \ldots, s$ are real numbers and a, b, c, d, e, g are not all zero. Then the tangent plane to this surface at (x_0, y_0, z_0) is $(ax_0 + dy_0 + ez_0 + h)(x - x_0) + (by_0 + dx_0 + gz_0 + p)(y - y_0) + (cz_0 + ex_0 + gy_0 + q)(z - z_0) = ax_0x + by_0y + cz_0z + d(xy_0 + x_0y) + e(xz_0 + x_0z) + g(yz_0 + z_0y) + h(x + x_0) + p(y + y_0) + q(z + z_0) + s - [ax_0^2 + by_0^2 + cz_0^2 + 2(dx_0y_0 + ex_0z_0 + gy_0z_0) + 2(hx_0 + py_0 + qz_0) + s]$. Since (x_0, y_0, z_0) is on the quadric, this equation reduces to

$$(13) \quad axx_0 + byy_0 + czz_0 + d(xy_0 + x_0y) + e(xz_0 + x_0z) \\ + g(yz_0 + y_0z) + h(x + x_0) + p(y + y_0) + q(z + z_0) + s = 0.$$

This formula arises from the original formula (12) of a quadric surface by the replacement of square terms such as x^2 by corresponding products such as xx_0, terms $2xy$ by $xy_0 + yx_0$, and terms $2x$ by $x + x_0$. It may then be easily remembered.

ILLUSTRATIVE EXAMPLE

Find an equation of the tangent plane to the quadric surface $3x^2 + 4y^2 - 2z^2 + 6xy - 3xz + 6x - 3z - 7 = 0$ at the point $(-1, 3, 2)$.

Solution

The equation is $-3x + 12y - 4z + 3(3x - y) - \frac{3}{2}(2x - z) + 3(x - 1) - \frac{3}{2}(z + 2) - 7 \equiv 6x + 9y - 4z - 13 = 0$.

EXERCISE

Write an equation of the tangent plane to each of the following quadrics at the given points:

(a) $x^2 - 2y^2 + 5z^2 = 18$, $(0, -1, 2)$
(b) $3x^2 + 4y^2 + 2z = 0$, $(0, 1, -2)$
(c) $x^2 - xy + z^2 = 3$, $(1, -1, -1)$
(d) $2xy - 3xz + 5z = 2$, $(1, -1, 2)$
(e) $2x^2 - y^2 + z^2 + 2xy + 4xz + x = 0$, $(0, -1, 1)$

CHAPTER 4
SPHERES

1. Equations of spheres. Let r be a positive real number and $P_0 = (x_0, y_0, z_0)$ be a fixed point. Then we define a *sphere* S of radius r and center at P_0 to be the locus of all points P whose distance from P_0 is r. It follows that a point P is on S if and only if $\sqrt{(x - x_0)^2 + (y - y_0)^2 + (z - z_0)^2} = r$, that is, if and only if P is a solution of the equation

$$(1) \qquad (x - x_0)^2 + (y - y_0)^2 + (z - z_0)^2 = r^2.$$

We have then succeeded in deriving an equation of an arbitrary sphere. Evidently a sphere is a quadric surface.

Theorem 1. *Let* a, b, c, d, e *be real numbers such that* $a \neq 0$ *and put*

$$(2) \quad x_0 = \frac{-b}{x}, \quad y_0 = \frac{-c}{a}, \quad z_0 = \frac{-d}{a}, \quad \rho = \frac{b^2 + c^2 + d^2 - ae}{a^2}.$$

Then the equation

$$(3) \quad f(x, y, z) = a(x^2 + y^2 + z^2) + 2(bx + cy + dz) + e = 0$$

has no real points as solutions if $\rho < 0$ *and is called an equation of an* **imaginary** *sphere. If* $\rho = 0$, *the only solution of* $f(x, y, z) = 0$ *is the point* $P_0 = (x_0, y_0, z_0)$ *and the equation is said to be an equation of a* **point** *sphere. The only remaining case is the case* $\rho > 0$ *and* $f(x, y, z) = 0$ *is an equation of the sphere whose center is* P_0 *and whose radius is* $\sqrt{\rho}$.

For a point is a solution of $f(x, y, z) = 0$ if and only if it is a solution of $(1/a)f(x, y, z) = 0$, that is, of

$$x^2 + y^2 + z^2 + \frac{2b}{a}x + \frac{2c}{a}y + \frac{2d}{a}z = \frac{-e}{a}.$$

This equation is equivalent, in view of the definitions of formula (2), to

$$(4) \qquad (x - x_0)^2 + (y - y_0)^2 + (z - z_0)^2 = \rho.$$

There are no real solutions (x, y, z) of this equation if $\rho < 0$,

54

its only real solution is (x_0, y_0, z_0) if $\rho = 0$, and it reduces to formula (1) with $r^2 = \rho$ if $\rho > 0$.

EXERCISES

1. Write an equation of a sphere with center at P_0 and radius r in the following cases:

(a) $P_0 = (0, 0, 0)$, $r = 5$ (d) $P_0 = (1, 6, 2)$, $r = 6$
(b) $P_0 = (-1, 2, 0)$, $r = 3$ (e) $P_0 = (1, 2, 2)$, $r = 3$
(c) $P_0 = (-1, -1, 3)$, $r = 7$ (f) $P_0 = (-2, 6, 3)$, $r = 7$

2. Determine the center and radius of the following spheres:
(a) $x^2 + y^2 + z^2 + 6x + 4y + 8z + 25 = 0$
(b) $x^2 + y^2 + z^2 + 2x - 6y + 4z + 10 = 0$
(c) $2(x^2 + y^2 + z^2) + 2x - 3y + 2z - 1 = 0$
(d) $x^2 + y^2 + z^2 - 3x + 5y - 7z = 0$
(e) $3(x^2 + y^2 + z^2) + x + y + z = 6$

3. Show that the lines passing through the center of a sphere are normal lines to the sphere.

2. Spheres satisfying given conditions. The equation of formula (3) is a linear homogeneous function of the five parameters (variables not regarded as point coordinates) a, b, c, d, e. It may then be made to satisfy four conditions such as passing through four points.

ILLUSTRATIVE EXAMPLES

I. Find an equation of the sphere with center at $(-1, 2, 3)$ and passing through the point $(1, -1, 2)$.

Solution

The restriction on the center is a set of three conditions. We use the formula (1) and write $(x + 1)^2 + (y - 2)^2 + (z - 3)^2 = r^2$. Then

$$(1 + 1)^2 + (1 - 2)^2 + (2 - 3)^2 = 4 + 9 + 1 = 14 = r^2.$$

Ans. $(x + 1)^2 + (y - 2)^2 + (z - 3)^2 = 14$.

II. Find an equation of the sphere through the points $(1, -1, 0)$, $(2, 1, 1)$, $(3, -1, 4)$, $(-1, -1, 2)$.

Solution

We substitute the coordinates of the four given points to obtain the equations

$$2a + 2(b - c) + e = 0$$
$$6a + 2(2b + c + d) + e = 0$$
$$26a + 2(3b - c + 4d) + e = 0$$
$$6a + 2(-b - c + 2d) + e = 0$$

Then $e = -2(a + b - c)$, and we subtract the first of our equations from the remaining equations and divide by 2 to obtain

$$\begin{aligned} 2a + b + 2c + d &= 0 \\ 12a + 2b \quad + 4d &= 0 \\ 2a - 2b \quad + 2d &= 0 \end{aligned}$$

This gives $2c = -(2a + b + d)$, and we need to solve

$$\begin{aligned} 6a + b &= -2d \\ a - b &= -d \end{aligned}$$

The solution is $7a = -3d$, and we avoid fractions by selecting $d = -14$, $a = 6$, $b = a + d = -8$, $2c = -(12 - 8 - 14) = 10$, $c = 5$, $e = -2(6 - 8 - 5) = 14$. Thus an equation is $6(x^2 + y^2 + z^2) + 2(-8x + 5y - 14z) + 14 = 0$.

$Ans.$ $3(x^2 + y^2 + z^2) - 8x + 5y - 14z + 7 = 0$.

EXERCISES

1. Find an equation of a sphere with center at P_0 and passing through P_1 in the following cases:

(a) $P_0 = (0, 0, 0), P_1 = (-1, 2, 3)$
(b) $P_0 = (-1, 1, 0), P_1 = (1, -1, 2)$
(c) $P_0 = (1, -1, 3), P_1 = (2, 3, -4)$
(d) $P_0 = (-1, -3, 2), P_1 = (-2, 1, -4)$
(e) $P_0 = (0, -1, 0), P_1 = (4, 0, -3)$
(f) $P_0 = (1, 2, 3), P_1 = (-1, 0, 2)$

2. Find an equation of a sphere passing through the points P_0, P_1 of Exercise 1 and the points P_2, P_3 as follows:

(a) $P_2 = (1, -2, 1), P_3 = (1, 1, 2)$
(b) $P_2 = (1, 2, -1), P_3 = (0, 0, 1)$
(c) $P_2 = (2, 1, 1), P_3 = (1, 0, -3)$
(d) $P_2 = (1, -1, 2), P_3 = (2, 3, -4)$
(e) $P_2 = (-2, -1, 1), P_3 = (-1, 1, 1)$
(f) $P_2 = (3, 2, 1), P_1 = (2, 2, -1)$

3. Linear families of spheres.   Consider the two spheres S_1 and S_2 defined by the equations

(5) $f(x, y, z) \equiv a_1(x^2 + y^2 + z^2) + 2(b_1x + c_1y + d_1z) + e_1 = 0$,
(6) $g(x, y, z) \equiv a_2(x^2 + y^2 + z^2) + 2(b_2x + c_2y + d_2z) + e_2 = 0$.

Then the equation

(7) $sf(x, y, z) + tg(x, y, z) = 0$

is an equation of the form of formula (3) in which

(8) $a = sa_1 + ta_2$, $b = sb_1 + tb_2$, $c = sc_1 + tc_2$,
$d = sd_1 + td_2$, $e = se_1 + te_2$.

It follows that if

(9) $$a = sa_1 + ta_2 \neq 0$$

the equation of formula (7) is an equation of a (point, imaginary or ordinary) sphere. We shall therefore call the family of surfaces defined by formula (7), as s and t range over all real numbers, a *family of spheres generated* by the two given spheres.

If $P_0 = (x_0, y_0, z_0)$ is any point on *both* $f(x, y, z) = 0$ and $g(x, y, z) = 0$, then P_0 is also on the surface $sf + tg = 0$. Thus all spheres of the family generated by the two given spheres contain all of points of intersection of these two spheres. Conversely, if S is a sphere through the curve C of intersection of two spheres S_1 and S_2, and S contains a point P_0 not on C, then S is the sphere $S(s, t)$ of the family defined by formula (7) when we take $s = g(x_0, y_0, z_0)$ and $t = -f(x_0, y_0, z_0)$. For there is at most one sphere through the curve C and a point not on C. The solution may, of course, be an imaginary sphere or a point sphere.

The values of s and t for which $a = sa_1 + ta_2 = 0$ yield the member of the family of spheres whose equation is

(10) $$2a_2(b_1x + c_1y + d_1z) - 2a_1(b_2x + c_2y + d_2z) \\ + (a_2e_1 - a_1e_2) = 0.$$

This is the equation of a plane except when $a_2(b_1, c_1, d_1) = a_1(b_2, c_2, d_2)$. The plane is called the *radical plane* of the two spheres. It contains the curve C of intersection of the two spheres and C is then the curve of intersection of the radical plane with either sphere. It follows that if two spheres intersect in at least two distinct points they intersect in a circle in their radical plane.

EXERCISES

1. Show that the line joining the centers of two intersecting spheres cuts their radical plane in the center of the circle of intersection.

2. Find the center and radius of the circle of intersection of the sphere $x^2 + y^2 + z^2 = 4$ and the sphere $(x - 1)^2 + y^2 + z^2 = 9$.

3. Find an equation of the sphere through the circle of intersection of the spheres of Exercise 2 and the following points determining the center and radius of the solution sphere in all cases where it is not a point or an imaginary sphere.

(a) $(-1, 2, 3)$

(b) $(4, 0, 0)$

(c) $(-3, 0, 0)$

(d) $(1, -1, 2)$

(e) $(-1, 1, -3)$

(f) $(1, 1, 1)$

(g) $(-1, -1, 0)$

(h) $(2, -1, 3)$

4. Angles between spheres. The tangent planes to the spheres of formulas (5) and (6) at a point of intersection (x_0, y_0, z_0) are given by

$$(11) \quad a_1(xx_0 + yy_0 + zz_0) + b_1(x + x_0) + c_1(y + y_0) \\ + d_1(z + z_0) + e_1 = 0,$$

$$(12) \quad a_2(xx_0 + yy_0 + zz_0) + b_2(x + x_0) + c_2(y + y_0) \\ + d_2(z + z_0) + e_2 = 0.$$

Then the corresponding sets of direction numbers are $(a_1x_0 + b_1, a_1y_0 + c_1, a_1z_0 + d_1)$ and $(a_2x_0 + b_2, a_2y_0 + c_2, a_2z_0 + d_2)$. The square of the length of the first of these vectors is

$$(a_1x_0 + b_1)^2 + (a_1y_0 + c_1)^2 + (a_1z_0 + d_1)^2 = a_1^2(x_0^2 + y_0^2 \\ + z_0^2) + 2a_1(b_1x_0 + c_1y_0 + d_1z_0) + b_1^2 + c_1^2 + d_1^2 \\ = b_1^2 + c_1^2 + d_1^2 - a_1e_1,$$

since (x_0, y_0, z_0) is a point of $f_1(x, y, z) = 0$. Similarly the square of the length of the second vector is $b_2^2 + c_2^2 + d_2^2 - a_2e_2$. The inner product of the two vectors is

$$g = (a_1x_0 + b_1)(a_2x_0 + b_2) + (a_1y_0 + c_1)(a_2y_0 + c_2) \\ + (a_1z_0 + d_1)(a_2z_0 + d_2) = a_1a_2(x_0^2 + y_0^2 + z_0^2) \\ + (a_1b_2 + a_2b_1)x_0 + (a_1c_2 + a_2c_1)y_0 + (a_1d_2 + a_2d_1)z_0 \\ + b_1b_2 + c_1c_2 + d_1d_2 = \tfrac{1}{2}a_2[a_1(x_0^2 + y_0^2 + z_0^2) \\ + 2(b_1x_0 + c_1y_0 + d_1z_0)] + \tfrac{1}{2}a_1[a_2(x_0^2 + y_0^2 + z_0^2) \\ + 2(b_2x_0 + c_2y_0 + d_2z_0)] + b_1b_2 + c_1c_2 + d_1d_2.$$

Then $g = -\tfrac{1}{2}(a_2e_1 + a_1e_2) + b_1b_2 + c_1c_2 + d_1d_2$ and the cosine of the angle between two planes at any point of their circle of intersection is given by the formula

(13)

$$\cos\theta = \frac{|2(b_1b_2 + c_1c_2 + d_1d_2) - (a_2e_1 + a_1e_2)|}{2\sqrt{(b_1^2 + c_1^2 + d_1^2 - a_1e_1)(b_2^2 + c_2^2 + d_2^2 - a_2e_2)}}$$

Since this formula does not involve the coordinates x_0, y_0, z_0, we have proved the following result:

Theorem 2. *Two spheres intersect at the same angle at all points of their circle of intersection.*

CHAPTER 5
QUADRIC SURFACES

1. Ellipsoids. In this chapter we shall discuss the geometric properties of the quadric surfaces given by a special set of equations. This discussion will yield the properties of all quadrics, since we shall prove in Chap. 7 that any equation of a quadric can be carried into one of the equations we shall discuss, by a suitable choice of coordinate axes.

If a, b, c are any positive real numbers, the equation

(1)
$$\frac{x^2}{a^2} + \frac{y^2}{b^2} + \frac{z^2}{c^2} = 1$$

is an equation of a surface called an *ellipsoid*. Such a surface is symmetrical with respect to the origin, the coordinate planes, and the coordinate axes, since it is a function of x^2, y^2, and z^2.

The six points $(\pm a, 0, 0)$, $(0, \pm b, 0)$, $(0, 0, \pm c)$ are called the *vertices* of the ellipsoid. They are the points of intersection of the coordinate axes and the surface. The three line segments joining the pairs of vertices on each coordinate axis are called the *axes* of the ellipsoid. The axes intersect at the origin of coordinates, and we call this point the *center* of the ellipsoid. The line segments joining the center to the vertices have lengths a, b, c and these lengths are called the *semiaxes* of the ellipsoid. If we order these three positive real numbers $2a, 2b, 2c$, the largest of the corresponding segments is called the *major axis* of the ellipsoid, the next largest the *mean axis*, and the smallest the *minor axis*.

We shall show in Chap. 7 that *every plane section of a quadric is a conic*. In particular we shall show that the plane sections of an ellipsoid are ellipses. The plane sections by the planes $z = k$ are ellipses

(2)
$$\frac{x^2}{a^2} + \frac{y^2}{b^2} = 1 - \frac{k^2}{c^2} = \frac{c^2 - k^2}{c^2}$$

provided that $c^2 > k^2$. If $c^2 = k^2$, the corresponding planes

$z = \pm c$ are tangent to the ellipsoid, and there are no real points of intersection if $k > c$. The student should make a similar analysis of the plane sections by planes $x = k$ and $y = k$ and should examine Fig. 11 carefully.

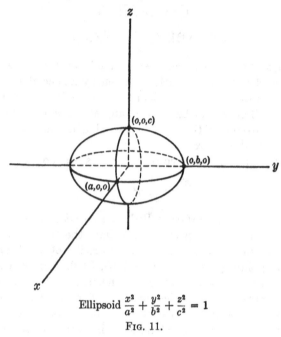

Ellipsoid $\dfrac{x^2}{a^2} + \dfrac{y^2}{b^2} + \dfrac{z^2}{c^2} = 1$

Fig. 11.

If two of the numbers a, b, c are equal, the ellipsoid is a surface of revolution about that axis which corresponds to the remaining letter; for example, the ellipsoid

$$(3) \qquad \frac{x^2 + y^2}{a^2} + \frac{z^2}{c^2} = 1$$

is an ellipsoid obtained by revolving the ellipse

$$(4) \qquad \frac{x^2}{a^2} + \frac{z^2}{c^2} = 1, \qquad y = 0$$

about the z axis. If $a > c$, this ellipsoid of revolution is the result of a revolution of an ellipse about its minor axis and is called an *oblate spheroid*. If $a < c$, the ellipsoid is the result of a revolution about its major axis and is called a *prolate spheroid*. If $a = c$, the ellipsoid of revolution is a sphere.

If $\alpha \neq 0$, $\beta \neq 0$, $\gamma \neq 0$, d, e, g, h are real numbers, the equation

(5) $f(x, y, z) \equiv \alpha x^2 + \beta y^2 + \gamma z^2 + (dx + ey + gz) + h = 0$

is equivalent to

(6) $\alpha \left(x + \dfrac{d}{2\alpha} \right)^2 + \beta \left(y + \dfrac{e}{2\beta} \right)^2 + \gamma \left(z + \dfrac{g}{2\gamma} \right)^2 = \rho,$

where

(7) $\rho = -h + \dfrac{d^2}{4\alpha} + \dfrac{e^2}{4\beta} + \dfrac{g^2}{4\gamma}.$

Suppose that α, β, γ all have the same sign. Then if $\rho = 0$, the only real point satisfying $f(x, y, z) = 0$ is the point (x_0, y_0, z_0) defined by

(8) $\qquad x_0 = -\dfrac{d}{2\alpha}, \qquad y_0 = -\dfrac{e}{2\beta}, \qquad z_0 = -\dfrac{g}{2\gamma},$

and the equation is called an equation of a *point ellipsoid*. If ρ and α have opposite signs, there are no real solutions of $f(x, y, z) = 0$ and this is an equation of what is called an *imaginary ellipsoid*. If α and ρ have the same signs, the equation $f(x, y, z) = 0$ is clearly an equation of the ellipsoid whose semiaxes are

(9) $\qquad\qquad \sqrt{\dfrac{\rho}{\alpha}}, \qquad \sqrt{\dfrac{\rho}{\beta}}, \qquad \sqrt{\dfrac{\rho}{\gamma}},$

whose axes are on the lines of intersection of the planes $x = x_0$, $y = y_0$, $z = z_0$, and whose center is the point (x_0, y_0, z_0).

EXERCISES

1. Give the coordinate of the center, the lengths of the semiaxes, and the equations of the lines on which the axes lie for the following ellipsoids:

(a) $9x^2 + 36y^2 + 4z^2 - 18x + 144y + 117 = 0$
(b) $x^2 + 2y^2 + 3z^2 - 2x + 8y - 6z + 11 = 0$
(c) $3x^2 + 4y^2 + 5z^2 + 6x - 16y + 10z + 23 = 0$
(d) $3x^2 + 3y^2 + 2z^2 - 6x - 12y - 12z + 29 = 0$

2. Give an equation of an ellipsoid with axes parallel to the coordinate axes, major axis $2a$ parallel to the y axis, minor axis $2c$ parallel to the z axis, and center at P in the following cases:

(a) $P = (-1, 2, 3)$, $a = 4$, $b = 3$, $c = 1$
(b) $P = (1, -2, 0)$, $a = 5$, $b = \sqrt{5}$, $c = 2$
(c) $P = (-1, -3, 1)$, $a = 2$, $b = \sqrt{3}$, $c = \sqrt{2}$
(d) $P = (4, -1, 2)$, $a = 3$, $b = 2$, $c = \sqrt{2}$

3. Give the equations of Exercise 2 if the major axis is parallel to the z axis and the minor axis to the x axis.

2. Quadric cones. Every set of positive real numbers a, b, c defines an equation

$$(10) \qquad \frac{x^2}{a^2} + \frac{y^2}{b^2} - \frac{z^2}{c^2} = 0,$$

which is homogeneous in x, y, z and thus is an equation of a quadric cone with vertex at the origin. It is evidently a surface symmetric with respect to the origin, the coordinate axes, and the coordinate planes.

The nature of such a surface is indicated by its sections with planes $z = k \neq 0$. These sections are the curves

$$(11) \qquad z = k, \qquad \frac{x^2}{a^2} + \frac{y^2}{b^2} = \frac{k^2}{c^2},$$

and thus are ellipses with semiaxes

$$(12) \qquad \frac{ah}{c}, \qquad \frac{bh}{c},$$

where $h = |k|$. The student should analyze the sections made by the planes $x = k$ and $y = k$.

Every plane through the origin and not parallel to the x axis is a plane $x = \alpha y + \beta z$, where α and β are real. The intersection of such a plane with the cone consists of points satisfying the equation $x = \alpha y + \beta z$ and the equation

$$(13) \qquad \frac{(\alpha y + \beta z)^2}{a^2} + \frac{y^2}{b^2} - \frac{z^2}{c^2} = 0.$$

The coefficient of y^2 is a positive real number, and we may designate it by

$$(14) \qquad \gamma^2 = \frac{\alpha^2}{a^2} + \frac{1}{b^2}.$$

The coefficient of yz may then be designated by $2\gamma\delta$ and that of z^2 by $\delta^2 - \rho$. Then formula (13) becomes

$$(15) \qquad (\gamma y + \delta z)^2 - \rho z^2 = 0.$$

If $\rho < 0$, the only points on the given plane and cone are the points $z = \gamma y + \delta z = 0$, $x = \alpha y + \beta z$, and thus $\gamma y = 0$, $y = 0$, $x = 0$. Hence, when $\rho < 0$, the only point of intersection of the plane and the cone is the origin. When $\rho \geqq 0$, the intersection consists of the line

(16) $x = \alpha y + \beta z, \qquad \gamma y + (\delta + \sqrt{\rho})z = 0$

and the line

(17) $x = \alpha y + \beta z, \qquad \gamma y + (\delta - \sqrt{\rho})z = 0.$

These lines coincide if $\rho = 0$.

Every plane through the origin is either the plane $z = 0$ which cuts the cone of formula (10) in the origin or is a plane which is surely not parallel either to the x axis or to the y axis. Since x and y have symmetric roles in formula (10) if a plane is parallel to the x axis and therefore not parallel to the y axis, we may repeat the argument above with the roles of x and y interchanged. We have therefore proved the following theorem:

Theorem 1. *The intersection of the quadric cone of formula* (10) *with a plane through its vertex consists either of the vertex or of two lines that may be coincident.*

3. Hyperboloids. If the numbers α, β, γ of formula (5) are not zero but do not all have the same sign, we may multiply by -1 if necessary and hence assume that

(18) $\alpha > 0, \qquad \beta > 0, \qquad \gamma < 0.$

We may then convert formula (5) into formula (6) and see that if $\rho = 0$ formula (6) becomes the equation

(19) $\dfrac{(x - x_0)^2}{a^2} + \dfrac{(y - y_0)^2}{b^2} - \dfrac{(z - z_0)^2}{c^2} = 0$

of a quadric cone with

$$\alpha = \frac{1}{a^2}, \qquad \beta = \frac{1}{b^2}, \qquad -\gamma = \frac{1}{c^2}$$

and vertex (x_0, y_0, z_0) given by formula (8). Assume then that $\rho \neq 0$.

If $\rho > 0$, formula (6) becomes the equation

(20) $\dfrac{(x - x_0)^2}{a^2} + \dfrac{(y - y_0)^2}{b^2} - \dfrac{(z - z_0)^2}{c^2} = 1,$

where

(21) $a^2 = \dfrac{\rho}{\alpha}, \qquad b^2 = \dfrac{\rho}{\beta}, \qquad c^2 = -\dfrac{\rho}{\gamma}.$

This is an equation of a surface called a *hyperboloid of one sheet.* The point x_0, y_0, z_0 is called the *center* of this surface, and the surface is symmetric with respect to the translated origin, the translated coordinate axes, and the translated coordinate planes

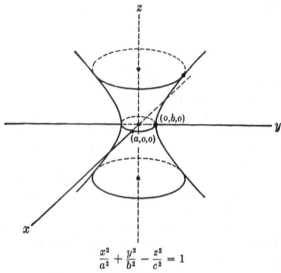

$$\frac{x^2}{a^2} + \frac{y^2}{b^2} - \frac{z^2}{c^2} = 1$$

FIG. 12.   Hyperboloid of one sheet.

after a translation of axes that carries the origin to the center. Let us assume then that the origin is the center and therefore study the equation

(22) $$\frac{x^2}{a^2} + \frac{y^2}{b^2} - \frac{z^2}{c^2} = 1.$$

The sections of this hyperboloid by the plane $z = k$ are ellipses

(23) $$\frac{x^2}{a^2} + \frac{y^2}{b^2} = \frac{k^2 + c^2}{c^2}, \qquad z = k$$

with center at the origin for all real values of k.   The semiaxes of these ellipses are

(24) $$\frac{a}{c} \sqrt{k^2 + c^2}, \qquad \frac{b}{c} \sqrt{k^2 + c^2}$$

and increase as $|k|$ increases.

The sections of the hyperboloid of formula (22) by planes $y = k$ are hyperbolas

(25)
$$\frac{x^2}{a^2} - \frac{z^2}{c^2} = \frac{b^2 - k^2}{b^2}$$

for all values of k except $k = \pm b$, and the semiaxes of these hyperbolas are

(26)
$$\frac{a}{b}\sqrt{|b^2 - k^2|}, \qquad \frac{c}{b}\sqrt{|b^2 - k^2|}.$$

When $k = \pm b$, the corresponding sections are pairs of straight lines

(27)
$$z = \pm \frac{c}{a}x, \qquad y = k.$$

A similar analysis of the plane sections by planes $x = k$ should be carried out by the student.

When $\rho < 0$, formula (6) becomes

(28)
$$\frac{(x - x_0)^2}{a^2} + \frac{(y - y_0)^2}{b^2} - \frac{(z - z_0)^2}{c^2} = -1,$$

where

(29)
$$a^2 = -\frac{\rho}{\alpha}, \qquad b^2 = -\frac{\rho}{\beta}, \qquad c^2 = \frac{\rho}{\gamma}.$$

This surface is called a *hyperboloid of two sheets*. The point

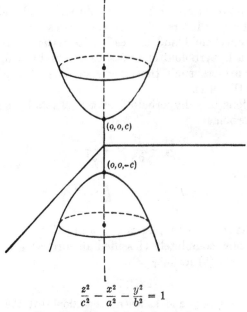

$$\frac{z^2}{c^2} - \frac{x^2}{a^2} - \frac{y^2}{b^2} = 1$$

Fig. 13. Hyperboloid of two sheets.

(x_0, y_0, z_0) is called the *center* of the hyperboloid, and we may translate the axes to the center and thus simplify the equation. Let us assume then that the origin is the center and therefore study the equation

$$(30) \qquad \frac{z^2}{c^2} - \frac{x^2}{a^2} - \frac{y^2}{b^2} = 1.$$

We shall call the numbers a, b, c of formulas (19), (20), and (28) the semiaxes of the three types of surfaces we are considering.

The sections of a hyperboloid of two sheets by the planes $z = k$ are given by

$$(31) \qquad \frac{x^2}{a^2} + \frac{y^2}{b^2} = \frac{k^2 - c^2}{c^2},$$

and therefore are ellipses with semiaxes

$$(32) \qquad \frac{a \sqrt{k^2 - c^2}}{c}, \qquad \frac{b \sqrt{k^2 - c^2}}{c}$$

provided that $|k| > c$. If, however, $-c < k < c$, the plane $z = k$ does not intersect the hyperboloid and thus there are no points on the hyperboloid of formula (30) between the plane $z = c$ and the plane $z = -c$. The points $(0, 0, c)$ and $(0, 0, -c)$ are the respective intersections of the planes $z = c$ and $z = -c$ with the hyperboloid and are called its *vertices*. It should be noted that a hyperboloid of two sheets actually consists of two separated surfaces such that $z \geqq c$ on one of the parts and $z \leqq -c$ on the other.

The sections of a hyperboloid of two sheets by planes $x = k$ are all hyperbolas

$$(33) \qquad \frac{z^2}{c^2} - \frac{y^2}{b^2} = \frac{k^2 + a^2}{a^2}$$

with semiaxes

$$(34) \qquad \frac{c \sqrt{k^2 + a^2}}{a}, \qquad \frac{b \sqrt{k^2 + a^2}}{a}.$$

Similarly the sections by planes $y = k$ are all hyperbolas.

We have now completely classified all surfaces given by equations of formula (5) for $\alpha\beta\gamma \neq 0$.

EXERCISES

1. Give a necessary and sufficient condition that the surfaces of formulas (19), (20), and (28) shall be surfaces of revolution.

2. Classify the following surfaces and give the center, the axes of symmetry, and the semiaxes in each case:

(a) $x^2 + 2y^2 - z^2 + 6x - 4y + 11 = 0$

(b) $x^2 - 2y^2 - 3z^2 + 4x - 6y + 12z + 11 = 0$

(c) $2x^2 - 3y^2 + 4z^2 + 6x - 9y + 12z = 0$

(d) $2x^2 - 2y^2 - 3z^2 + 16y + 12z = 44$

(e) $4x^2 + y^2 - 9z^2 + 8x + 2y - 18z + 11 = 0$

(f) $4x^2 - 4y^2 - 9z^2 - 8x + 8y - 36z = 0$

4. Lines on a hyperboloid. We shall begin our discussion of lines on a hyperboloid by proving the following:

Theorem 2. *A hyperboloid of two sheets contains no lines.*

For the line $x = x_0 + t\xi$, $y = y_0 + t\eta$, $z = z_0 + t\zeta$ will lie on the hyperboloid

$$\frac{x^2}{a^2} - \frac{y^2}{b^2} - \frac{z^2}{c^2} = 1$$

if and only if

$$(35) \qquad \frac{(x_0 + t\xi)^2}{a^2} = \frac{(y_0 + t\eta)^2}{b^2} + \frac{(z_0 + t\zeta)^2}{c^2} + 1$$

for all values of t. This equation is then an identity in t and the coefficients of 1, t, and t^2 in the two members must be equal, *i.e.*,

$$(36) \qquad \frac{x_0^2}{a^2} = \frac{y_0^2}{b^2} + \frac{z_0^2}{c^2} + 1, \qquad \frac{x_0\xi}{a^2} = \frac{y_0\eta}{b^2} + \frac{z_0\zeta}{c^2}, \qquad \frac{\xi^2}{a^2} = \frac{\eta^2}{b^2} + \frac{\zeta^2}{c^2}.$$

Then

$$(37) \qquad \left(\frac{x_0\xi}{a^2}\right)^2 = \left(\frac{y_0\eta}{b^2} + \frac{z_0\zeta}{c^2}\right)^2 = \left(\frac{\eta^2}{b^2} + \frac{\zeta^2}{c^2}\right)\left(\frac{y_0^2}{b^2} + \frac{z_0^2}{c^2} + 1\right),$$

and

$$(38) \qquad \frac{\eta^2}{b^2} + \frac{\zeta^2}{c^2} + \frac{\eta^2 z_0^2 + \zeta^2 y_0^2 - 2y_0\eta z_0\zeta}{(bc)^2} = 0.$$

However, this expression is a sum of three real squares and can vanish only when the squares vanish separately. Thus $\eta = \zeta = 0$, and the last equation of formula (36) implies that $\xi^2 = 0$, $\xi = 0$. Then our line degenerates to $x = x_0$, $y = y_0$, $z = z_0$, that is, no line lies wholly on the surface.

We next pass on to the case of a hyperboloid of one sheet. The lines

$$(39) \qquad \frac{x}{a} + \frac{z}{c} = \lambda\left(1 + \frac{y}{b}\right), \qquad \lambda\left(\frac{x}{a} - \frac{z}{c}\right) = 1 - \frac{y}{b},$$

defined for all real numbers λ, consist of points wholly on the surface defined by formula (22). For if $P_0 = (x_0, y_0, z_0)$ is a point on a line of formula (39), then

$$(40) \quad \frac{x_0^2}{a^2} - \frac{z_0^2}{c^2} = \lambda\left(1 + \frac{y_0}{b}\right)\left(\frac{x_0}{a} - \frac{z_0}{c}\right) = \left(1 + \frac{y_0}{b}\right)\left(1 - \frac{y_0}{b}\right)$$
$$= 1 - \frac{y_0^2}{b^2},$$

and P_0 is on the hyperboloid. Conversely, if P_0 is on the surface, then

$$(41) \quad \left(\frac{x_0}{a} + \frac{z_0}{c}\right)\left(\frac{x_0}{a} - \frac{z_0}{c}\right) = \left(1 + \frac{y_0}{b}\right)\left(1 - \frac{y_0}{b}\right).$$

If the denominator of

$$(42) \quad \lambda = \frac{1 - y_0/b}{(x_0/a) - (z_0/c)}$$

is not zero, the point P_0 lies on the unique line of the family of formula (39) determined by this value of λ. If both numerator and denominator are zero in formula (42) P_0 is on the line uniquely determined by $\lambda = x_0/a$. Finally, if the numerator of formula (42) is not zero and the denominator is zero, then P_0 is on the line

$$(43) \quad 1 + \frac{y}{b} = 0, \qquad \frac{x}{a} - \frac{z}{c} = 0.$$

This line is *regarded* as being that member of the family of lines of formula (39) defined for the infinite value of λ.

The family of lines defined by formula (39) is called a *regulus* of the corresponding surface, and we have proved that through every point of the surface there passes one and only one line of the regulus. The family of lines defined by

$$(44) \quad \frac{x}{a} - \frac{z}{c} = \mu\left(1 + \frac{y}{b}\right), \qquad \mu\left(\frac{x}{a} + \frac{z}{c}\right) = 1 - \frac{y}{b}$$

is a second regulus of the surface of formula (22), and it should be evident that it can be similarly proved that one and only one line of this second regulus passes through a point on the surface.

The two reguli of a hyperboloid have no lines in common; for let us suppose that a particular value of λ and a particular value of μ give the same line. The point $y = b$, $x = a\lambda$, $z = c\lambda$ is on the line of formula (39) and when substituted in formula (44) gives $\mu = 0$. Then the point $y = b$, $x = z = 0$ is on this second line and substitution in formula (39) yields $\lambda = 0$. However, the two resulting lines

$$(45) \qquad \frac{x}{a} = -\frac{z}{c}, \quad y = b; \qquad \frac{x}{a} = \frac{z}{c}, \quad y = b$$

have only the point $(0, b, 0)$ in common and cannot coincide.

If P is a point on a hyperboloid of one sheet, there are exactly two lines on the surface that pass through P. For the line through x_0, y_0, z_0 and having direction numbers ξ, η, ζ will lie on the surface S of formula (22) if and only if

$$\frac{(x_0 + t\xi)^2}{a^2} + \frac{(y_0 + t\eta)^2}{b^2} - \frac{(z_0 + t\zeta)^2}{c^2}$$

is identically zero in t. This requires that (x_0, y_0, z_0) shall be a point of S and that

$$\frac{\xi^2}{a^2} + \frac{\eta^2}{b^2} - \frac{\zeta^2}{c^2} = 0, \qquad \frac{x_0\xi}{a^2} + \frac{y_0\eta}{b^2} - \frac{z_0\zeta}{c^2} = 0.$$

Then (ξ, η, ζ) is a point not the origin and on the plane

$$\frac{x_0 x}{a^2} + \frac{y_0 y}{b^2} - \frac{z_0 z}{c^2} = 0,$$

and the cone

$$\frac{x^2}{a^2} + \frac{y^2}{b^2} - \frac{z^2}{c^2} = 0.$$

By Theorem 1 this plane through the origin cuts the cone in exactly two lines through the origin. Let the coordinate of two nonzero vectors on these two lines be (ξ_1, η_1, ζ_1) and (ξ_2, η_2, ζ_2), respectively, so that every point on the first line is a vector $(\xi, \eta, \zeta) = p(\xi_1, \eta_1, \zeta_1)$ and every point on the second line is a vector $(\xi, \eta, \zeta) = q(\xi_2, \eta_2, \zeta_2)$. Since all nonzero scalar multiples of a vector (ξ, η, ζ) define a set of direction numbers of the same line as (ξ, η, ζ), we have shown that there are not more than two lines through a point on the surface of a hyperboloid. We have also proved the existence of two lines lying in two reguli and the proof is complete.

EXERCISES

1. Substitute the value $y = b$ in formulas (39) and (44), and thus determine a point on a corresponding line of the reguli. Determine a

second point with $y = -b$, when $\lambda \neq 0$ and when $\mu \neq 0$. However, when $\lambda = 0$, show that $(a, b, -c)$ is a point of the line and that, when $\mu = 0$, (a, b, c) is a point of the line. Use these values and the infinite case to determine a set of direction numbers for any line of the two reguli.

2. By symmetry we see that a hyperboloid of formula (22) has the regulus

$$\frac{y}{b} + \frac{z}{c} = \xi \left(1 + \frac{x}{a} \right), \qquad \xi \left(\frac{y}{b} - \frac{z}{c} \right) = 1 - \frac{x}{a},$$

and the regulus

$$\frac{y}{b} - \frac{z}{c} = \eta \left(1 + \frac{x}{a} \right), \qquad \eta \left(\frac{y}{b} + \frac{z}{c} \right) = 1 - \frac{x}{a}.$$

By substituting the points of Exercise 1 show that the first of these two reguli is precisely the regulus of formula (44) and that the second is the regulus of formula (39).

3. Show that every plane through a line of the regulus of formula (39) also contains a line of the regulus of formula (44). HINT: The planes through a line of formula (39) are all of the form

$$\frac{x}{a} + \frac{z}{c} - \lambda \left(1 + \frac{y}{b} \right) = \mu \left[1 - \frac{y}{b} - \lambda \left(\frac{x}{a} - \frac{z}{c} \right) \right].$$

4. Show that every plane of Exercise 3 is tangent to the surface at the point of intersection of the two generators.

5. Write out the equations of the reguli of the following hyperboloids, and determine which pair of lines of the reguli pass through the corresponding point.

(a) $\dfrac{x^2}{4} + \dfrac{y^2}{9} - \dfrac{z^2}{16} = 1$, $(2, -1, \frac{4}{3})$

(b) $\dfrac{x^2}{9} + y^2 - \dfrac{z^2}{9} = 1$, $(3, \frac{1}{3}, -1)$

(c) $\dfrac{x^2 + y^2}{9} - z^2 = 1$, $(3\sqrt{2}, 0, 1)$

6. Show that there are no lines on the surface of an ellipsoid by using the method of the proof of Theorem 2.

5. Paraboloids. By consideration of symmetry we see that a discussion of formula (5) in the case where two of the numbers α, β, γ are not zero and the remaining number is zero need only be carried out for the case where $\alpha\beta \neq 0$, $\gamma = 0$. We may also assume that $\alpha > 0$, since we may multiply formula (5) by -1 if necessary. Let us also assume in this section that $g \neq 0$. We may then convert formula (5) into

(46) $\alpha(x - x_0)^2 + \beta(y - y_0)^2 + 2g(z - z_0) = 0,$

where

(47) $x_0 = -\dfrac{d}{2\alpha},$ $y_0 = -\dfrac{e}{2\beta},$ $z_0 = -\dfrac{h}{2g} + \dfrac{d^2}{8\alpha g} + \dfrac{e^2}{8\beta g}.$

We translate the axes to (x_0, y_0, z) and divide the equation of formula (46) by $-|2g|$ to convert it to the form

(48) $\dfrac{x^2}{a^2} + \dfrac{y^2}{b^2} = \epsilon z,$

if $\alpha\beta > 0$, and to

(49) $\dfrac{x^2}{a^2} - \dfrac{y^2}{b^2} = \epsilon z,$

if $\alpha\beta < 0$, where $\epsilon = 1$ or -1.

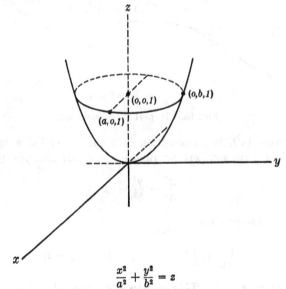

$$\frac{x^2}{a^2} + \frac{y^2}{b^2} = z$$

Fig. 14.   Elliptic paraboloid.

Equation (48) is an equation of a surface called an *elliptic paraboloid*. Its plane sections, by planes $z = k$, are ellipses

(50) $\dfrac{x^2}{a^2} + \dfrac{y^2}{b^2} = \epsilon k$

if $\epsilon k > 0$. The plane $z = 0$ cuts this parabola at the origin and this point is called the *vertex* of the paraboloid. The plane sections $x = k$ and $y = k$ of the paraboloid are parabolas.

(51) $$\frac{x^2}{a^2} = \epsilon z - \frac{k^2}{b^2}, \qquad y = k$$

and

(52) $$\frac{y^2}{b^2} = \epsilon z - \frac{k^2}{a^2}, \qquad x = k.$$

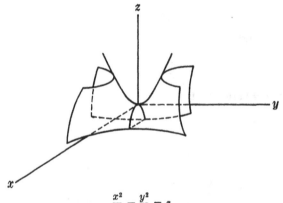

$$\frac{x^2}{a^2} - \frac{y^2}{b^2} = z$$

Fig. 15. Hyperbolic paraboloid.

Equation (49) is an equation of a surface called a *hyperbolic paraboloid*. Its sections, by planes $z = k$, are all hyperbolas

(53) $$\frac{x^2}{a^2} - \frac{y^2}{b^2} = \epsilon k$$

except for the section

$$\left(\frac{x}{a} + \frac{y}{b}\right)\left(\frac{x}{a} - \frac{y}{b}\right) = 0, \qquad z = 0$$

by the plane $z = 0$. This section consists of the line

$$\frac{x}{a} = \frac{y}{b}, \qquad z = 0$$

and the line

$$\frac{x}{a} = -\frac{y}{b}, \qquad z = 0.$$

The hyperbolic paraboloid of formula (49) has two reguli. They are given by the equations

(54) $$\frac{x}{a} + \frac{y}{b} = \lambda \epsilon z, \qquad \lambda \left(\frac{x}{a} - \frac{y}{b} \right) = 1$$

and

(55) $$\frac{x}{a} - \frac{y}{b} = \mu \epsilon z, \qquad \mu \left(\frac{x}{a} + \frac{y}{b} \right) = 1.$$

The student should show, as an exercise, that the lines of the first regulus are parallel to the plane $bx = ay$ and that those of the second regulus are parallel to the plane $bx = -ay$. These are then two distinct lines through each point of the surface. The hyperboloid of one sheet and the hyperbolic paraboloid are sometimes called *ruled surfaces*.

It is easy to show that there are no lines on an elliptic paraboloid; for a line is defined by direction numbers $(\xi, \eta, \zeta) \neq (0, 0, 0)$ and a point (x_0, y_0, z_0) such that

$$\phi(t) \equiv \frac{(x_0 + t\xi)^2}{a^2} + \frac{(y_0 + t\eta)^2}{b^2} - \epsilon(z_0 + t\zeta) = 0,$$

and thus

$$\frac{\xi^2}{a^2} + \frac{\eta^2}{b^2} = 0, \qquad 2\left(\frac{\xi}{a^2} + \frac{\eta}{b^2} \right) - \epsilon\zeta = 0.$$

Then $\xi = \eta = \zeta = 0$, a contradiction.

We have already found two distinct lines through each point (x_0, y_0, z_0) of the surface (49), and they are the line of formula (54) determined by

$$\lambda = \frac{(x_0/a) + (y_0/b)}{\epsilon z_0} = \frac{1}{(x_0/a) - (y_0/b)}.$$

and the line of formula (55) determined by

$$\mu = \frac{(x/a) - (y/b)}{\epsilon z_0} = \frac{1}{(x_0/a) + (y_0/b)}.$$

To prove that these are the only two lines, we examine the equation

$$\phi(t) \equiv \frac{(x_0 + \xi t)^2}{a^2} - \frac{(y_0 + \eta t)^2}{b^2} - \epsilon(z_0 + \zeta t) = 0$$

and thus see that

$$\frac{\xi^2}{a^2} = \frac{\eta^2}{b^2}, \qquad \frac{\xi x_0}{a^2} - \frac{\eta y_0}{b^2} - \epsilon\zeta = 0.$$

Then

$$\eta = \pm \frac{b}{a}\,\xi, \qquad \zeta = \left(\frac{x_0}{\epsilon a^2} \mp \frac{y_0}{\epsilon ab}\right)\xi,$$

i.e., all sets of direction numbers ξ, η, ζ are scalar multiples of the two sets

$$\left(1, \frac{b}{a}, \frac{x_0}{\epsilon a^2} - \frac{y_0}{ab}\right), \qquad \left(1, -\frac{b}{a}, \frac{x_0}{\epsilon a^2} + \frac{y_0}{\epsilon ab}\right).$$

This proves that there are exactly two lines on the surface through each point of a hyperbolic paraboloid.

EXERCISES

1. Determine a set of direction numbers for the lines of formulas (54) and (55) by using the value $z = 0$ and the value $y = kb$.

2. Write out the equations of the reguli of the following paraboloids and of the particular lines of the reguli through the corresponding points:

(a) $\dfrac{x^2}{4} - \dfrac{y^2}{9} = 2z$, $(4, 3, \frac{3}{2})$

(b) $\dfrac{x^2}{4} - y^2 = -2z$, $(2, 1, 0)$

(c) $\dfrac{x^2}{9} - \dfrac{y^2}{4} = z$, $(6, 6, -5)$

6. Cylinders. The equation of formula (5) reduces to

$$(56) \qquad \alpha\left(x + \frac{d}{2\alpha}\right)^2 + \beta\left(y + \frac{d}{2\beta}\right)^2 = \rho$$

if $\alpha\beta \neq 0$ and $g = 0$. If $\alpha > 0$, $\beta > 0$, this equation is an equation of what is called an *elliptic cylinder*. Its sections by planes $z = k$ are evidently ellipses, and we have already proved that there is one line on the surface through each point of it.

If $\alpha > 0$, $\beta > 0$, and $\rho = 0$, the surface defined by formula (56) consists only of the point

$$x = -\frac{d}{2\alpha}, \qquad y = -\frac{d}{2\beta},$$

and if $\alpha\beta > 0$, $\rho < 0$, there are no real points on the surface.

If $\alpha > 0$ and $\beta < 0$, formula (56) is an equation of a hyperbolic cylinder when $\rho = 0$ and of two planes through the z axis when $\rho = 0$. This completes the case of surfaces defined by formula (56).

There remains the case where two of the numbers α, β, γ of formula (5) are zero. We may then assume that $\alpha = 1$, since we can always divide the equation by $\alpha \neq 0$. Then formula (5) becomes

$$(57) \qquad x^2 + 2(dx + ey + gz) + h = 0.$$

In Chap. 7, we shall show how to select a coordinate system such that a surface defined by formula (57) has the form of (57) with $g = 0$. Then the equation reduces to

$$(58) \qquad (x + d)^2 = -2ey + d^2 - h.$$

This is an equation of a parabolic cylinder when $e \neq 0$; it is an equation of two distinct parallel planes when $e = 0$, $d^2 \neq h$; it is an equation $x = -d$ of two coincident planes when $e = 0$, $d^2 = h$; and is a surface with no real points when $e = 0$, $d < h$.

7. Classification of quadric surfaces. A quadric surface is defined by an equation $f(x, y, z) = 0$, where $f(x, y, z)$ is a polynomial of degree two. We shall prove in Chap. 7 that it is always possible to select a coordinate system such that the surface is defined by an equation of formula (5) where the numbers α, β, γ are independent of the choice of the coordinate system. Let us call these numbers a set of *characteristic roots* of the surface. They are unique apart from a proportionality factor t, which may be introduced by multiplying $f(x, y, z) = 0$ by $t \neq 0$.

Quadric surfaces may be classified according to the properties of their sets of characteristic roots and the number of lines of the surface through each point of the surface. The classification follows:

1. Characteristic roots all of the same sign. Such surfaces are either ellipsoids, point ellipsoids, or imaginary ellipsoids. There are no lines on such surfaces.

2. Three nonzero characteristic roots not all of the same sign.

a. The quadric cone. This is a surface containing a point (its vertex) through which pass infinitely many lines of the surface.

b. The hyperboloid of one sheet. There are precisely two lines of the surface through every point of it.

c. The hyperboloid of two sheets. There are no lines on this surface.

3. A ZERO ROOT AND TWO NONZERO ROOTS HAVING THE SAME SIGN.

a. The elliptic paraboloid. There are no lines on this surface.

b. The elliptic cylinder. There is exactly one line of the surface through each point of it. This surface may reduce to a point or it may be imaginary.

4. A ZERO ROOT AND TWO NONZERO ROOTS HAVING OPPOSITE SIGNS.

a. The hyperbolic paraboloid. There are two distinct lines of the surface S through each point of S.

b. The hyperbolic cylinder. One and only one line of S passes through each point of S.

c. Two distinct planes through a line. There are infinitely many lines of S through each point of S.

5. THERE IS ONLY ONE NONZERO CHARACTERISTIC ROOT.

a. The parabolic cylinder. One and only one line of S passes through each point of S.

b. Two distinct parallel planes. There are infinitely many lines on S through each point of S.

c. Two coincident planes. This is the only place in the classification where lines and sets of characteristic roots are not adequate to separate cases. It is evident that the line criterion separates (*c*) and (*a*) into distinct types. The separation of (*b*) and (*c*) requires a count of planes rather than of lines.

CHAPTER 6
THEORY OF MATRICES

1. Matrices. A rectangular array is called a *matrix*. An m by n matrix is an array

$$(1) \qquad A = \begin{pmatrix} a_{11} & \cdots & a_{1n} \\ \cdot & \cdots & \cdot \\ a_{m1} & \cdots & a_{mn} \end{pmatrix}.$$

The horizontal lines in this array are called its *rows*. They are n-dimensional vectors and the ith row is $(a_{i1}, \ldots, a_{in})$. Every (row) vector is thus a one by n matrix.

The vertical lines in A are called its *columns*. The jth column of A is

$$(2) \qquad \begin{pmatrix} a_{1j} \\ \cdot \\ \cdot \\ \cdot \\ a_{mj} \end{pmatrix},$$

and A has n columns. Each column may be thought of as being an m-dimensional column (*i.e.*, vertically written) vector, and is an m by one matrix.

The scalar in the ith row and jth column of A has been designated, by implication, as a_{ij}, where the first subscript always will indicate the label of the row and the second subscript the label of the column in which a_{ij} appears. It will be convenient to use the notation

$$(3) \qquad A = (a_{ij}) \quad (i = 1, \ldots, m; j = 1, \ldots, n),$$

instead of the more cumbersome notation of formula (1).

A matrix A is called a *square* matrix if it has as many rows as columns, that is, A is an n by n matrix. A square matrix having n rows and columns is called an *n-rowed* square matrix or a square matrix of *order n*.

77

ORAL EXERCISES

1. Read off the elements a_{13}, a_{24}, a_{51}, a_{26}, a_{36}, a_{42}, a_{55} in the following five by six matrix:

$$\begin{pmatrix} 2 & -1 & 3 & 1 & 4 & 5 \\ 0 & -2 & -4 & 6 & 7 & 8 \\ 1 & 0 & -2 & 4 & 3 & -5 \\ -1 & 2 & 1 & -2 & 1 & 1 \\ 0 & 0 & 2 & 0 & -1 & 0 \end{pmatrix}$$

2. Read off the third row and the fourth column of this matrix.

2. Addition and scalar multiplication. The *sum* $A + B$ of two rectangular matrices A and B is defined only when A and B have the same size. If

(4) $$A = (a_{ij}), \qquad B = (b_{ij})$$
$$(i = 1, \ldots, m; j = 1, \ldots, n)$$

then the sum of the m by n matrix A and the m by n matrix B is the m by n matrix

(5) $$C = A + B = (c_{ij}), \qquad c_{ij} = a_{ij} + b_{ij}$$
$$(i = 1, \ldots, m; j = 1, \ldots, n).$$

In words, *matrices are added by adding corresponding elements.*

Lemmas 1 to 5 of Chap. 1 are special cases of the corresponding properties of matrix addition. The student should formulate and verify the matrix properties. Note that the zero m by n matrix is the matrix whose elements are all zero. We shall use the symbol 0 for such a matrix no matter what its size is, and the size will always be given by the context. The matrix $-A$ is the matrix whose elements are the negatives of the elements of A; $B - A$ is obtained by subtracting the elements of A from those of B.

If a is a number and A is an m by n matrix, the *scalar product* is the m by n matrix obtained by multiplying *every* element of A by a. Thus if $A = (a_{ij})$, the element in the ith row and jth column of aA is aa_{ij}. Clearly $1A = A$, $(-1)A = -A$, $aA = 0$ if $a = 0$. We also have the properties

(6) $a(bA) = (ab)A, \qquad (a + b)A = aA + bA,$
$$a(A + B) = aA + aB$$

for all numbers a, b and all m by n matrices A, B. The verification of these properties is very simple and will not be given here.

ORAL EXERCISES

1. Form the matrix sums $A + B$ in the following cases:

(a) $A = \begin{pmatrix} 3 & -1 & 2 & 4 \\ 2 & 1 & 0 & -3 \\ 4 & -2 & 1 & 0 \end{pmatrix}$, $\quad B = \begin{pmatrix} 1 & 2 & -3 & -4 \\ -1 & -2 & 1 & 3 \\ -2 & 2 & -1 & 0 \end{pmatrix}$

(b) $A = \begin{pmatrix} 2 & 1 & 3 \\ 4 & 1 & 2 \\ -2 & 1 & 0 \\ 0 & -1 & 3 \end{pmatrix}$, $\quad B = \begin{pmatrix} 3 & -1 & -2 \\ -3 & 1 & -1 \\ 2 & -1 & 0 \\ 1 & 1 & -3 \end{pmatrix}$

2. Give $-A$ and $-B$ for the matrices A and B of Oral Exercise 1.

3. Give $2A$ and $-3B$ for the matrices A and B of Oral Exercise 1.

4. Give $-2A + B$ for the matrices A and B of Oral Exercise 1.

3. Matrix multiplication. The product AB of two rectangular matrices A and B is defined *only* if the number of columns of A is equal to the number of rows of B.

Let us then assume that A is an m by n matrix and that B is an n by t matrix. Then the ith row of A is an n-dimensional vector

$$(7) \qquad A_i = (a_{i1}, a_{i2}, \ldots, a_{in}).$$

The kth column of B is also an n-dimensional (column) vector, which we may write horizontally as

$$(8) \qquad B^{(k)} = (b_{1k}, b_{2k}, \ldots, b_{nk}).$$

The inner product

$$(9) \qquad g_{ik} = A_i \cdot B^{(k)} = a_{i1}b_{1k} + \cdots + a_{in}b_{nk}$$

of the ith row of A and the kth column of B is defined for $i = 1, \ldots, m$ and $k = 1, \ldots, t$, and therefore the m by t matrix

$$G = (g_{ik}) \quad (i = 1, \ldots, m; k = 1, \ldots, t)$$

is also defined. We call G the *product* of A and B and write

$$G = AB.$$

It is a matrix whose element in the ith row and kth column is the inner product of the ith row of A and the kth column of B.

Lemma 1. *Matrix multiplication is associative, that is,* (AB)C = A(BC) *for all* m *by* n *matrices* A, n *by* t *matrices* B, *and* t *by* s *matrices* C.

For we may write $A = (a_{ij})$, $B = (b_{jk})$, $C = (c_{kp})$, where $i = 1$, $\ldots$, m; $j = 1$, $\ldots$, n; $k = 1$, $\ldots$, t; $p = 1$, $\ldots$, s. The general element of AB is $\displaystyle\sum_{j=1}^{n} a_{ij}b_{jk}$ and that of $(AB)C$ is $d_{ip} = \displaystyle\sum_{k=1}^{t}\left(\sum_{j=1}^{n} a_{ij}b_{jk}\right)c_{kp}$. The general element of BC is $\displaystyle\sum_{k=1}^{t} b_{jk}c_{kp}$ and that of $A(BC)$ is $g_{ip} = \displaystyle\sum_{j=1}^{n} a_{ij}\left(\sum_{k=1}^{t} b_{jk}c_{kp}\right)$. The two finite double sums d_{ip} and g_{ip} are sums of exactly the same products $a_{ij}b_{jk}c_{kp}$ and are equal for all values of $i = 1$, $\ldots$, m, and of $p = 1$, $\ldots$, s. This proves that $A(BC) = (AB)C$.

Lemma 2. *Matrix multiplication is distributive with addition,* i.e.,

(10) $A(B + C) = AB + AC$, $(B + C)D = BD + CD$

for all m *by* n *matrices* A, n *by* t *matrices* B *and* C, *and* t *by* s *matrices* D.

For let $A = (a_{ij})$, $B = (b_{jk})$, $C = (c_{jk})$, where $i = 1$, $\ldots$, m; $j = 1$, $\ldots$, n; $k = 1$, $\ldots$, t. Then the element in the ith row and kth column of $A(B + C)$ is $\displaystyle\sum_{j=1}^{n} a_{ij}(b_{jk} + c_{jk})$ and the matrix equality $A(B + C) = AB + AC$ is equivalent to the formula

(11) $$\sum_{j=1}^{n} a_{ij}(b_{jk} + c_{jk}) = \sum_{j=1}^{n} a_{ij}b_{jk} + \sum_{j=1}^{n} a_{ij}c_{jk}.$$

This formula is evidently correct. The proof that $(B + C)D = BD + CD$ is carried out similarly.

EXERCISES

1. Form the matrix product AB in the following cases:

(a) $A = \begin{pmatrix} 2 & -1 & 3 \\ 1 & 2 & -1 \\ 3 & 0 & 2 \\ 4 & 0 & -3 \end{pmatrix}$, $B = \begin{pmatrix} 2 & 1 & 3 & 4 \\ -1 & 2 & 0 & 0 \\ 3 & -1 & 2 & -3 \end{pmatrix}$

(b) $A = \begin{pmatrix} 1 & 3 \\ 2 & -1 \\ 0 & 2 \end{pmatrix}$, $B = \begin{pmatrix} 1 & 2 & -1 & 3 \\ 3 & -1 & 2 & -1 \end{pmatrix}$

(c) $A = \begin{pmatrix} -1 & 2 & 1 & 0 \\ 2 & -1 & 1 & 3 \\ -1 & 1 & -1 & 1 \end{pmatrix}$, $B = \begin{pmatrix} -1 & 4 & 0 \\ 2 & -1 & 0 \\ -3 & 2 & 2 \\ 1 & 0 & -1 \end{pmatrix}$

2. Form the matrix product BA in those cases of Exercise 1 where it exists.

3. Form the matrix products $A(BC)$ and $(AB)C$ if A and B are given by part (b) of Exercise 1 and

$$C = \begin{pmatrix} 1 & -1 \\ 2 & 0 \\ -1 & 2 \\ 3 & -1 \end{pmatrix}.$$

4. Form $A(B + C)$ and AB, AC, $AB + AC$ if A and B are given by part (c) of Exercise 1 and

$$C = \begin{pmatrix} 1 & -3 & 0 \\ -2 & 0 & -1 \\ 2 & -1 & -1 \\ -1 & 1 & 2 \end{pmatrix}.$$

4. Transposition. If the rows and columns of an m by n matrix A are interchanged, the result is an n by m matrix A^* (read A transpose) called the *transpose* of A. The operation of transposition thus consists of writing the column vectors of A as row vectors of A^*. If a_{ij} is the element in the ith row and jth column of A, then a_{ij} is the element in the jth row and ith column of A^*.

If $P = (x_1, \ldots, x_n)$ and $Q = (y_1, \ldots, y_n)$, the matrix product PQ^* of the one by n matrix P and the n by one matrix Q^* is given by

$$(x_1, \ldots, x_n) \begin{pmatrix} y_1 \\ \cdot \\ \cdot \\ \cdot \\ y_n \end{pmatrix} = x_1 y_1 + \cdots + x_n y_n.$$

Hence, the inner product $P \cdot Q$ of two n-dimensional vectors P and Q coincides with the matrix product PQ^*.

Theorem 1. *The transpose of a product of matrices is the product of the transposes of the factors in reverse order, that is,*

$$(12) \qquad\qquad (AB)^* = B^*A^*.$$

For let $A = (a_{ij})$ and $B = (b_{jk})$, where $i = 1, \ldots, m$; $j = 1, \ldots, n; k = 1, \ldots, t$. Then the element in the kth row and ith column of $(AB)^*$ is $c_{ik} = \sum\limits_{j=1}^{n} a_{ij}b_{jk}$. The element in the kth row and jth column of B^* is b_{jk}, the element in the jth row and ith column of A^* is a_{ij}, and therefore the element in the kth row and ith column of B^*A^* is $\sum\limits_{j=1}^{n} b_{jk}a_{ij} = c_{ij}$. Hence, $(AB)^* = B^*A^*$.

ORAL EXERCISE

Show that if A is an m by n matrix then $C = AA^* = C^*$.

5. Special matrices. The elements a_{ii} of any matrix $A = (a_{ij})$ are called the *diagonal* elements of A and the line of these elements is called the *diagonal* of A. The elements a_{ij} with $j > i$ (column label greater than row label) are said to lie *above the diagonal* of A and those with $j < i$ are said to lie *below the diagonal* of A.

A matrix is called a *triangular* matrix if A is a square matrix such that either all of the elements above the diagonal in A are zero or all of the elements below the diagonal in A are zero. Examples are

$$\begin{pmatrix} 2 & -1 & 4 & 3 \\ 0 & 2 & -1 & 5 \\ 0 & 0 & 6 & 2 \\ 0 & 0 & 0 & -1 \end{pmatrix}, \qquad \begin{pmatrix} 1 & 0 & 0 & 0 & 0 \\ 2 & 6 & 0 & 0 & 0 \\ -1 & 5 & 0 & 0 & 0 \\ 4 & 1 & 2 & 3 & 0 \\ 5 & -1 & 6 & 7 & -8 \end{pmatrix}.$$

If $A = (a_{ij})$ is a square matrix such that $a_{ij} = 0$ for every $i \neq j$, we call A a *diagonal* matrix. The notation

(13) $$A = diag\ \{a_1, a_2, \ldots, a_n\}$$

will be used for an n-rowed diagonal matrix whose ith diagonal element a_{ii} is the number a_i.

A diagonal matrix whose diagonal elements are all equal is called a *scalar* matrix. The scalar matrix whose diagonal elements are all unity is called the *identity* matrix and will usually be designated simply by I. Then we may verify by direct multiplication

(14) $$I_m A = A I_n$$

for any m by n matrix A where we have used subscripts on I to indicate the sizes of the identity matrices involved. Observe that a scalar matrix whose diagonal elements are all a is the scalar product aI and that the matrix product $(aI)A$ is equal to the scalar product aA for any matrix A.

6. Products in terms of rows and columns. Let $A = (a_{ij})$ and $B = (b_{jk})$, where $i = 1, \ldots , m; \; j = 1, \ldots , n; \; k = 1, \ldots , t$. Designate the jth row of B by

$$B_j = (b_{j1}, b_{j2}, \ldots , b_{jn}) \qquad (j = 1, \ldots , n)$$

and form

$$C_i = a_{i1}B_1 + a_{i2}B_2 + \cdots + a_{in}B_n.$$

Then the element in the kth column of the vector C_i is $a_{i1}b_{1k} + a_{i2}b_{2k} + \cdots + a_{in}b_{nk}$. This is the element in the ith row and kth column of AB, and we have proved the result about rows in the following theorem. The result about columns is proved similarly.

Theorem 2. *The ith row of* AB *is that linear combination of the rows of* B *whose coefficients make up the ith row of* A. *The kth column of* AB *is that linear combination of the columns of* A *whose coefficients make up the kth column of* B.

We apply Theorem 2 in the case where A is a diagonal matrix so that each row of A has only one nonzero element and this element is in the ith column of A. This yields the following result:

Theorem 3. *Let* $A = \text{diag} \{a_1, \ldots , a_n\}$ *and* B *be an* n *by* n *matrix. Then the ith row of* AB *is* a_i *times the ith row of* B. *If* $D = \text{diag} \{d_1, \ldots , d_n\}$ *the jth column of* BD *is* d_j *times the jth column of* B.

The result given in formula (14) is clearly a special case of Theorem 3.

EXERCISES

1. Show that if A is a square matrix such that $AD = DA$, where D is a diagonal matrix having distinct diagonal elements, then A is a diagonal matrix.

2. Use the property of Exercise 1 to prove that if A is an n-rowed square matrix commutative with all n-rowed square matrices (*i.e*, having the property that $AB = BA$ for every B) then A is a scalar matrix. HINT: Form the products AC_j and C_jA where C_j has 1 in the first row and jth column and zeros elsewhere.

7. Partitioning of a matrix. The elements which appear in r of the rows and s of the columns of an m by n matrix A form an r by s matrix which is called a *submatrix* of A. The r rows need not be adjacent. When all r rows and all s columns of a submatrix of A are adjacent, then we shall refer to the submatrix as being a *block* of elements of A.

Every matrix A may be partitioned into four blocks, and we may write

$$(15) \qquad A = \begin{pmatrix} A_1 & A_2 \\ A_3 & A_4 \end{pmatrix}.$$

Here $A = (a_{ij})$ for $i = 1, \ldots, m$ and $j = 1, \ldots, n$ and the partitioning is completely determined when we write

$$(16) \qquad A_1 = (a_{ij}) \quad (i = 1, \ldots, r; j = 1, \ldots, s).$$

Then we are assuming that

$$
\begin{aligned}
A_2 &= (a_{ij}) & (i = 1, \ldots, r; j = s + 1, \ldots, n), \\
A_3 &= (a_{ij}) & (i = r + 1, \ldots, m; j = 1, \ldots, s), \\
A_4 &= (a_{ij}) & (i = r + 1, \ldots, m; j = s + 1, \ldots, n).
\end{aligned}
$$

Let A be partitioned as in formula (15). This partitioning is determined by the fact that A_1 is an r by s block in the first r rows and first s columns of A. Then we shall say that an n by t matrix B is partitioned similarly to A if

$$(17) \qquad B = \begin{pmatrix} B_1 & B_2 \\ B_3 & B_4 \end{pmatrix}$$

where B_1 has s rows. The number q of columns in B_1 is completely arbitrary and so is the number of rows in A_1. We then multiply A by B and have

$$(18) \qquad AB = \begin{pmatrix} A_1B_1 + A_2B_3 & A_1B_2 + A_2B_4 \\ A_3B_1 + A_4B_3 & A_3B_2 + A_4B_4 \end{pmatrix}.$$

Formula (18) states that if two matrices are partitioned similarly we may multiply them as if they were two-rowed square matrices whose elements are, of course, not numbers but blocks. The formula is very easy to derive. We first note that formula (18) states that the block of AB which makes up its first r rows and first q columns is $A_1B_1 + A_2B_3$. The first r rows of A make up a block of elements that may be designated by (A_1A_2) and the first q columns of B may be designated by

$$\begin{pmatrix} B_1 \\ B_3 \end{pmatrix}.$$

The element in the ith row and kth column of AB is $\sum\limits_{j=1}^{n} a_{ij}b_{jk} =$

$\sum\limits_{j=1}^{s} a_{ij}b_{jk} + \sum\limits_{j=s+1}^{n} a_{ij}b_{jk}$. If $i = 1, \ldots, r$ and $k = 1, \ldots, q$ the

sum $\sum\limits_{j=1}^{n} a_{ij}b_{jk}$ is the element in the ith row and kth column of

$A_1 B_1$ and $\sum\limits_{j=s+1}^{n} a_{ij}b_{jk}$ is the element in the ith row and kth column

of $A_2 B_3$. This proves that

$$(A_1 \quad A_2)\begin{pmatrix} B_1 \\ B_3 \end{pmatrix} = A_1 B_1 + A_2 B_3.$$

The remaining relations in formula (18) are proved similarly.

It should be observed that formula (18) is the result of *focusing attention* on certain rows of A and certain columns of B and of

breaking up the general sum $\sum\limits_{j=1}^{n} a_{ij}b_{jk}$ into the sum of two partial

sums $\sum\limits_{j=1}^{s} a_{ij}b_{jk}$ and $\sum\limits_{j=s+1}^{n} a_{ij}b_{jk}$. The formula is of particular use

in cases where a block is a zero matrix or an identity matrix. For example, let A and B be n-rowed square matrices, and write

$$(19) \qquad A = \begin{pmatrix} I & C \\ 0 & I \end{pmatrix}, \qquad B = \begin{pmatrix} B_1 & B_2 \\ B_3 & B_4 \end{pmatrix}.$$

In this formula we are using the same symbol I for an r-rowed identity matrix and for an $(n - r)$-rowed identity matrix, the matrix C is an r by $n - r$ matrix, and the matrix 0 is an $n - r$ by r zero matrix. Then

$$(20) \qquad AB = \begin{pmatrix} B_1 + CB_3 & B_2 + CB_4 \\ B_3 & B_4 \end{pmatrix}.$$

It follows that in this case the last $n - r$ rows of AB coincide with the corresponding rows of B. However, each of the first r

rows of AB is the sum of the corresponding row of B and a linear combination of the last r rows of B.

EXERCISES

1. Show that if

$$\begin{pmatrix} A_1 & A_2 \\ A_3 & A_4 \end{pmatrix} \begin{pmatrix} I & 0 \\ C & I \end{pmatrix} = 0$$

then A_2 and A_4 are zero matrices.

2. Show that if

$$\begin{pmatrix} A_1 & 0 \\ A_2 & A_4 \end{pmatrix} \begin{pmatrix} B_1 & 0 \\ 0 & B_2 \end{pmatrix} = I$$

then A_1B_1 and A_4B_2 are identity matrices. This will be shown later to imply that $A_1B_1 = B_1A_1$. Use this property to prove that $A_2 = 0$.

8. Determinants. Every n-rowed square matrix $A = (a_{ij})$ has a *determinant* that is a certain function of the elements of A. We designate the determinant of A by $|A|$, by $|a_{ij}|$, or by

(21)
$$\begin{vmatrix} a_{11} & a_{12} & \cdots & a_{1n} \\ a_{21} & a_{22} & \cdots & a_{2n} \\ \cdots & \cdots & \cdots & \cdots \\ a_{n1} & a_{n2} & \cdots & a_{nn} \end{vmatrix}.$$

We shall refer to the determinant of an n-rowed square matrix as an n-*rowed determinant* or a determinant of *order* n. Rectangular matrices with $m \neq n$ do not have determinants, but every r-rowed square submatrix of any matrix A has a determinant that is called an r-*rowed minor* of A.

We shall define $|A|$ by an induction on n. If $n = 1$, the matrix $A = (a_{11})$, and we define $|A| = a_{11}$. Assume then that we have a definition of all $(n - 1)$-rowed determinants.

Each element a_{ij} of an n-rowed square matrix A defines a row and a column of A. Delete this row and column of A, and obtain a submatrix A_{ij} of A. This submatrix A_{ij} has $n - 1$ rows and columns and thus has a determinant $|A_{ij}|$ that is an $(n - 1)$-rowed minor of A defined for every i and j. Designate by b_{ji} (*not by* b_{ij}) the number

(22)
$$(-1)^{i+j}|A_{ij}|$$

and call this signed minor the *cofactor* of a_{ij} in A. Then b_{ji} is the element in the jth row and ith column of a matrix

$$(23) \qquad\qquad \text{adj } A = (b_{ji}) \qquad (j, i = 1, \ldots, n),$$

which is called the *adjoint* of A.

Let us now form $A(\text{adj } A) = C = (c_{ik})$, where

$$(24) \qquad\qquad c_{ik} = a_{i1}b_{1k} + \cdots + a_{in}b_{nk}$$

is obtained by multiplying the elements of the ith row of A by the cofactors of the elements of the kth row of A and adding the resulting products. We similarly form $(\text{adj } A)A = D = (d_{ki})$, where

$$(25) \qquad\qquad d_{ki} = b_{k1}a_{1i} + \cdots + b_{kn}a_{ni}$$

is obtained by multiplying the elements in the ith column of A by the cofactors of the elements in the kth column and adding the resulting products. In particular c_{ii} is the sum of the products of elements in the ith row of A by their cofactors and d_{kk} is the sum of the products of the elements in the kth column of A by their cofactors.

It can be shown that the n numbers c_{ii} and the n numbers d_{kk} are all equal. *We define the common value of these $2n$ numbers to be the determinant of A.* We have thus given not only a definition of a determinant but what are called the *expansions* of it according to any row and any column.

EXERCISES

1. Expand the following determinants according to the first row:

$$(a) \begin{vmatrix} 1 & -1 & 2 \\ 3 & 1 & -2 \\ 1 & 2 & 3 \end{vmatrix} \qquad (b) \begin{vmatrix} 2 & 1 & -2 \\ 0 & -1 & 1 \\ -2 & 1 & 0 \end{vmatrix} \qquad (c) \begin{vmatrix} 2 & -1 & 1 & 1 \\ 1 & 1 & 0 & 1 \\ 2 & 2 & 0 & -1 \\ 1 & -3 & -2 & 1 \end{vmatrix}$$

$$(d) \begin{vmatrix} 2 & -1 & -2 & 1 \\ -1 & 2 & 1 & 1 \\ 1 & -3 & -1 & -2 \\ 2 & 1 & -2 & 3 \end{vmatrix} \qquad (e) \begin{vmatrix} 3 & -1 & 2 & -1 \\ -1 & 1 & 0 & 1 \\ 1 & 2 & 3 & 2 \\ 2 & -3 & -1 & 1 \end{vmatrix}$$

2. Expand each of the determinants of Exercise 1 according to the second row.

3. Compute the adjoint of each of the following matrices:

$$(a) \begin{pmatrix} -1 & 2 & 1 \\ 2 & -1 & 3 \\ 1 & 2 & -1 \end{pmatrix} \qquad (b) \begin{pmatrix} 1 & -1 & 1 \\ 2 & 2 & -1 \\ -1 & 1 & 3 \end{pmatrix} \qquad (c) \begin{pmatrix} 0 & -1 & 2 \\ 2 & -1 & 3 \\ -4 & 2 & 1 \end{pmatrix}$$

9. Properties of determinants. We shall assume the following properties of determinants.

Lemma 3. *A square matrix and its transpose have the same determinant.*

Lemma 4. *Let* B *be the result of interchanging two rows (columns) of a square matrix* A. *Then* $|\text{B}| = -|\text{A}|$.

Lemma 5. *Let* B *be the result of multiplying a row (column) of* A *by a number* a. *Then* $|\text{B}| = a|\text{A}|$.

Note that this result implies that $|aA| = a^n|A|$.

Lemma 6. *Let* B *be the result of adding a scalar multiple of a row (column) of* A *to another row (column). Then* $|\text{B}| = |\text{A}|$.

Lemma 7. *Let a row (column) of* A *be a scalar multiple of another row (column). Then* $|\text{A}| = 0$.

If we multiply the elements of the ith row (column) of A by the cofactors of the elements in its kth row (column), the result is the expansion according to the kth row of the determinant of a matrix B obtained by replacing the kth row (column) of A by its ith row (column). Then two rows (columns) of B are equal and $|B| = 0$. But $|B|$ is the number c_{ik} of formula (24) in the row case and is the number d_{ki} of formula (25) in the column case. Hence, $c_{ik} = c_{ki} = 0$ for $i \neq k$ and $c_{ii} = d_{ii} = |A|$. This yields the following result:

Theorem 4. *Let* A *be an* n-*rowed square matrix. Then*

$$\text{A}(adj\ \text{A}) = (adj\ \text{A})\text{A} = |\text{A}|\text{I}$$

is the scalar product of the n-*rowed identity matrix* I *by the determinant of* A.

Any square matrix A can be converted into a diagonal matrix by a finite sequence of transformations of the types given in Lemmas 4 and 6. Indeed if $A \neq 0$, we can carry any nonzero element a_1 of A into its first row and column. We can then subtract multiples of the first row from the remaining rows and multiples of the first column from the remaining columns and convert our matrix into

$$B_1 = \begin{pmatrix} a_1 & 0 \\ 0 & A_1 \end{pmatrix},$$

where A_1 has $n - 1$ rows and columns. The same procedure may be applied to A_1 by applying transformations to the last $n - 1$ rows and columns of B_1, and we ultimately convert A into a diagonal matrix D. If A has been converted into D by transformations involving an even number of row and column interchanges, then $|A| = |D|$. Otherwise, we change the sign of a row before obtaining the final diagonal matrix D and will have $|A| = |D|$. Then $|A|$ may be computed by the use of

Lemma 8. *The determinant of a triangular matrix* T *is the product of the diagonal elements of* T.

It is clearly sufficient to consider the case where the elements above the diagonal in T are all zero. Then

$$T = \begin{vmatrix} a_1 & 0 \\ S_1 & T_1 \end{vmatrix}$$

where T_1 is a triangular matrix of $n - 1$ rows. Expand T according to its first row, and see that $|T| = a_1|T_1|$. Apply the same procedure to T_1, and ultimately obtain the result of the theorem.

We should observe that if the transformations used to convert A to a diagonal matrix D such that $|A| = |D|$ are applied to *any* square matrix B *of any size* the resulting matrix C will have the property that $|B| = |C|$. For C results from B by a finite number of transformations of the kind in Lemmas 4, 5, and 6 and there are only either an even number of those in Lemma 4 and none of those in Lemma 5 or there are an odd number of those in Lemma 4 and one of those in Lemma 5 with $a = -1$. We use this result to prove

Lemma 9. *Let*

(26) $$C = \begin{vmatrix} A & 0 \\ G & B \end{vmatrix}, \qquad D = \begin{vmatrix} A & H \\ 0 & B \end{vmatrix}$$

where A *and* B *are square matrices. Then*

(27) $$|C| = |D| = |A| \cdot |B|.$$

To prove this result, we apply a sequence of transformations to C and to D which convert A to a diagonal matrix A_0 and B to a diagonal matrix B_0 such that $|A| = |A_0|$ and $|B| = |B_0|$. These transformations convert C and D, respectively, to

$$C_0 = \begin{pmatrix} A_0 & 0 \\ G_0 & B_0 \end{pmatrix}, \qquad D_0 = \begin{pmatrix} A_0 & H_0 \\ 0 & B_0 \end{pmatrix}.$$

Then $|C_0| = |C|$, $|D_0| = |D|$. But the new matrices C_0 and D_0 are triangular matrices and their diagonal elements are the diagonal elements of A_0 and B_0. Then $|C_0| = |A_0| \cdot |B_0| = |D_0|$ and $|C| = |D| = |A| \cdot |B|$ as desired.

We use Lemma 9 in the proof of the very important result that we state as follows:

Theorem 5. *The determinant of a product of n-rowed square matrices is the product of the determinants of its factors, i.e.,* $|AB| = |A| \cdot |B|$.

To prove this result, we consider the $2n$-rowed square matrix

$$(28) \qquad C = \begin{pmatrix} A & 0 \\ -I & B \end{pmatrix}.$$

Form the product

$$(29) \qquad D = \begin{pmatrix} I & A \\ 0 & I \end{pmatrix} C = \begin{pmatrix} 0 & AB \\ -I & B \end{pmatrix},$$

as in formula (19). Only the first n rows of C have been altered and each of these rows is a corresponding row of C plus a linear combination of the last n rows of C. By Lemma 6, we have $|D| = |C|$. It requires n interchanges of rows and n changes of signs of rows to replace D by

$$(30) \qquad G = \begin{pmatrix} I & -B \\ 0 & AB \end{pmatrix}.$$

Then $|G| = (-1)^{2n}|D| = |C|$. By Lemma 9 we have $|G| = |I| \cdot |AB| = |AB|$ and by this same lemma $|C| = |A| \cdot |B|$. Hence, $|AB| = |A| \cdot |B|$ and our theorem is proved.

EXERCISES

1. Use the properties of determinants to reduce each determinant of Exercise 1 of Sect. 8 to triangular form and compute its value by the use of Lemma 8.

2. What are the determinants of the following matrices?

$$(a) \begin{pmatrix} -2 & 1 & 0 & 0 \\ 3 & -4 & 0 & 0 \\ 5 & 16 & 8 & 2 \\ 9 & 11 & 5 & 3 \end{pmatrix}, \qquad (b) \begin{pmatrix} 2 & 1 & 3 & 8 & 9 & -3 \\ 0 & 2 & 1 & 6 & 1 & 2 \\ 0 & 4 & 3 & 9 & 8 & 7 \\ 0 & 0 & 0 & 4 & 0 & 0 \\ 0 & 0 & 0 & -1 & 5 & 0 \\ 0 & 0 & 0 & 9 & 6 & 2 \end{pmatrix}.$$

10. The inverse of a matrix. A matrix A is said to be *nonsingular* if A is a square matrix and $|A| \neq 0$. A square matrix A whose determinant is zero is called a *singular* matrix.

If A is a nonsingular n-rowed matrix, we define a matrix

$$(31) \qquad A^{-1} = \frac{1}{|A|} \text{ adj } A$$

which we call the *inverse* of A. By Theorem 4

$$(32) \qquad AA^{-1} = A^{-1}A = I$$

is the n-rowed identity matrix. A singular matrix does not have an inverse. For if $AA^{-1} = I$, then $|AA^{-1}| = |A| \cdot |A^{-1}| = 1$ and $|A| \neq 0$. Note that

$$(33) \qquad |A^{-1}| = |A|^{-1}.$$

If A is nonsingular, the matrix equations

$$(34) \qquad AX = B, \qquad YA = C$$

have unique solutions X and Y for any given matrices B and C. Indeed if $AX = B$, then $A^{-1}(AX) = A^{-1}B = (A^{-1}A)X = IX = X$. Similarly if $YA = C$, then $CA^{-1} = (YA)A^{-1} = Y(AA^{-1}) = Y$. We have proved that the equations of formula (34) have the unique solutions

$$(35) \qquad X = A^{-1}B, \qquad Y = CA^{-1}.$$

In particular the equations $AX = I$ and $YA = I$ have the unique solution $X = Y = A^{-1}$. It follows that $AX = I$ if and only if $XA = I$, that is, $X = A^{-1}$.

The following results are almost trivial.

Theorem 6. *The inverse of a product is the product of the inverses of the factors in reverse order.*

For if $C = A_1 A_2 \cdots A_r$, then $CA_r^{-1}A_{r-1}^{-1} \cdots A_1^{-1} = I$, that is, $C^{-1} = A_r^{-1}A_{r-1}^{-1} \cdots A_1^{-1}$. In the simple case of two factors, we are stating that $ABB^{-1}A^{-1} = I$ and therefore $(AB)^{-1} = B^{-1}A^{-1}$.

Theorem 7. *The inverse of A^* is the transpose of A^{-1}.*

For $AA^{-1} = I$, $I^* = I = (A^{-1})^*A^*$ and therefore $(A^*)^{-1} = (A^{-1})^*$.

EXERCISE

Compute A^{-1} by using formula (31) in each of the following cases:

$$(a) \begin{pmatrix} 2 & -1 & 1 \\ 0 & 4 & 1 \\ 0 & 3 & 1 \end{pmatrix}, \qquad (b) \begin{pmatrix} 1 & 2 & 0 \\ 4 & 7 & 0 \\ 6 & -3 & 1 \end{pmatrix}, \qquad (c) \begin{pmatrix} 1 & 1 & 2 \\ 1 & 2 & 3 \\ 3 & -2 & 2 \end{pmatrix}.$$

11. Linear systems of equations. A system of m linear equations

$$
\begin{aligned}
a_{11}x_1 + a_{12}x_2 + \cdots + a_{1n}x_n &= y_1 \\
a_{21}x_1 + a_{22}x_2 + \cdots + a_{2n}x_n &= y_2 \\
&\cdots\cdots\cdots\cdots\cdots\cdots \\
a_{m1}x_1 + a_{m2}x_2 + \cdots + a_{mn}x_n &= y_m
\end{aligned}
\tag{36}
$$

in n unknowns $x_1, \ldots, x_n$ is called a *linear system*. If A is the m by n matrix, $A = (a_{ij})$ of the coefficients, and $P = (x_1, \ldots, x_n)$, $Q = (y_1, \ldots, y_m)$, we may write any linear system as an equivalent vector matrix equation

$$
AP^* = Q^*,
\tag{37}
$$

where

$$
P^* = \begin{pmatrix} x_1 \\ \cdot \\ \cdot \\ \cdot \\ x_n \end{pmatrix}, \qquad Q^* = \begin{pmatrix} y_1 \\ \cdot \\ \cdot \\ \cdot \\ y_m \end{pmatrix}.
$$

Any system of n equations in n unknowns with $|A| \neq 0$ may be solved by determinants or by elimination. Whatever procedure is used, the solution must be

$$
P^* = A^{-1}Q^*.
$$

It follows that to find the inverse of a given nonsingular matrix A we form the equations

$$
\begin{aligned}
a_{11}x_1 + \cdots + a_{1n}x_n &= y_1 \\
a_{21}x_1 + \cdots + a_{2n}x_n &= y_2 \\
&\cdots\cdots\cdots\cdots\cdots \\
a_{n1}x_1 + \cdots + a_{nn}x_n &= y_n
\end{aligned}
\tag{38}
$$

and solve these equations for $x_1, \ldots, x_n$ in terms of $y_1, \ldots, y_n$, thereby obtaining solutions of the form

$$
\begin{aligned}
x_1 &= b_{11}y_1 + \cdots + b_{1n}y_n \\
x_2 &= b_{21}y_1 + \cdots + b_{2n}y_n \\
&\cdots\cdots\cdots\cdots\cdots\cdots \\
x_n &= b_{n1}y_i + \cdots + b_{nn}y_n.
\end{aligned}
\tag{39}
$$

Then, if $B = (b_{ij})$, we see that $B = A^{-1}$ is the required inverse matrix.

ILLUSTRATIVE EXAMPLE

Compute the inverse of

$$A = \begin{pmatrix} 0 & 3 & 2 & 3 \\ 1 & 2 & 1 & 2 \\ -1 & 2 & 3 & 5 \\ -1 & 3 & 4 & 7 \end{pmatrix}.$$

Solution

We require the solution of the system of equations

$$\begin{aligned} 3x_2 + 2x_3 + 3x_4 &= y_1 \\ x_1 + 2x_2 + x_3 + 2x_4 &= y_2 \\ -x_1 + 2x_2 + 3x_3 + 5x_4 &= y_3 \\ -x_1 + 3x_2 + 4x_3 + 7x_4 &= y_4. \end{aligned}$$

We first eliminate x_1 and obtain

$$\begin{aligned} 4x_2 + 4x_3 + 7x_4 &= y_2 + y_3 \\ 5x_2 + 5x_3 + 9x_4 &= y_2 + y_4. \end{aligned}$$

Then $36x_4 - 35x_4 = x_4 = 4(y_2 + y_4) - 5(y_2 + y_3) = -y_2 - 5y_3 + 4y_4$. We also compute $x_2 + x_3 + 2x_4 = y_4 - y_3$ and have $3(y_4 - y_3) - y_1 = 3x_2 + 3x_3 + 6x_4 - (3x_2 + 2x_3 + 3x_4) = x_3 - 3x_4$, so that $x_3 = 3y_4 - 3y_3 - y_1 + 3y_2 + 15y_3 - 12y_4 = -y_1 + 3y_2 + 12y_3 - 9y_4$. We now obtain $x_2 = y_4 - y_3 - (-y_1 + 3y_2 + 12y_3 - 9y_4) - 2(-y_2 - 5y_3 + 4y_4) = y_1 - y_2 - 3y_3 + 2y_4$, $x_1 = -2(y_1 - y_2 - 3y_3 + 2y_4) - (-y_1 + 3y_2 + 12y_3 - 9y_4) - 2(-y_2 - 5y_3 + 4y_4) = -y_1 + y_2 + 4y_3 - 3y_4$. It follows that

$$A^{-1} = \begin{pmatrix} -1 & 1 & 4 & -3 \\ 1 & -1 & -3 & 2 \\ -1 & 3 & 12 & -9 \\ 0 & -1 & -5 & 4 \end{pmatrix}.$$

The result should be checked by computing $AA^{-1} = I$.

EXERCISES

1. Compute A^{-1} by the method above for the exercise of Sec. 11.

2. Compute A^{-1} for each of the following matrices:

(a) $\begin{pmatrix} -1 & 0 & 0 & -1 \\ 1 & 1 & 2 & 0 \\ 0 & 1 & 1 & -1 \\ 1 & 1 & 4 & -2 \end{pmatrix}$
(b) $\begin{pmatrix} -1 & 2 & -2 & 0 \\ 0 & -1 & 1 & 1 \\ -3 & 1 & -4 & 3 \\ 0 & 2 & -1 & -1 \end{pmatrix}$

(c) $\begin{pmatrix} -1 & 0 & -2 & 2 \\ -3 & 1 & -1 & -1 \\ 1 & 0 & 2 & -1 \\ -2 & 1 & 0 & 1 \end{pmatrix}$
(d) $\begin{pmatrix} -1 & 1 & 0 & 1 \\ 1 & 0 & 0 & 1 \\ -4 & 3 & -3 & 3 \\ 0 & -1 & 2 & 1 \end{pmatrix}$

12. Homogeneous systems. A linear system

(40)
$$a_{11}x_1 + \cdots + a_{1n}x_n = 0$$
$$a_{21}x_1 + \cdots + a_{2n}x_n = 0$$
$$\cdot \;\cdot\; \cdot \;\cdot\; \cdot \;\cdot\; \cdot \;\cdot\; \cdot \;\cdot\; \cdot \;\cdot\; \cdot$$
$$a_{m1}x_1 + \cdots + a_{mn}x_n = 0$$

is called a *homogeneous* system. It requires that

$$(41) \qquad\qquad AP^* = 0$$

and therefore that $P = (x_1, \ldots, x_n)$ be a vector such that

$$(42) \qquad\qquad A_i \cdot P = 0 \qquad\qquad (i = 1, \ldots, n),$$

where A_i is the ith row of P. Thus we see that a homogeneous system of equations proposes the *problem of finding an* n-*dimensional vector* P *which is orthogonal to* m *given* n-*dimensional vectors*. The zero vector is a trivial solution of the problem but we are usually interested in finding a nontrivial solution, *i.e.*, a nonzero vector P.

Theorem 8. *A homogeneous system of* m *equations in* n *unknowns has a nontrivial solution if* m < n.

A single equation $a_1x_1 + \cdots + a_mx_n = 0$ in at least two unknowns has a nontrivial solution. For we put $x_3 = \cdots = x_n = 0$ and see that if $a_1 = a_2 = 0$ the vector $(1, 1, 0, 0, \ldots, 0)$ is a nontrivial solution. But if one of a_1 and a_2 is not zero the vector $(-a_2, a_1, 0, \ldots, 0)$ is a nontrivial solution. We now make an induction on m and assume that the theorem is true for $m - 1$ equations in more than $m - 1$ unknowns.

The theorem is surely trivial if all coefficients are zero. Hence, we may assume that at least one coefficient is not zero. There is surely no loss of generality if we assume that this coefficient is a_{11}. This amounts to a relabeling of equations and variables if necessary. We may then subtract multiples $a_{11}^{-1}a_{i1}$ of the first equation from all the other equations and obtain our equivalent system. The first equation may be written as

$$(43) \qquad x_1 = -a_{11}^{-1}(a_{12}x_2 + \cdots + a_{1n}x_n),$$

and the remaining equations are free of x_1. We then have $m - 1$ equations in $n - 1$ unknowns. These equations are homogeneous and have a nontrivial solution $(x_2, \ldots, x_n)$. If x_1 is defined by formula (43), the vector $(x_1, \ldots, x_n)$ is a nontrivial solution of our original system.

Theorem 9. *A homogeneous system* $AP^* = 0$ *of* n *equations in* n *unknowns has a nontrivial solution* $P \neq 0$ *if and only if* $|A| = 0$.

For if $|A| \neq 0$, the equation $AP^* = 0$ implies that $A^{-1}(AP^*) = P^* = 0$, $P = 0$. Let us then suppose that $|A| = 0$. If $n = 1$, the system becomes $a_1x_1 = 0$, where $|A| = a_1 = 0$ and $x_1 \neq 0$ defines a nontrivial solution. Assume then that the theorem is true for the case of $n - 1$ equations in $n - 1$ unknowns. As in the proof of Theorem 8, we may assume that $a_{11} \neq 0$ and pass to an equivalent system with matrix

$$B = \begin{pmatrix} a_{11} & B_1 \\ 0 & B_2 \end{pmatrix},$$

where B_1 is an $(n - 1)$-dimensional vector, the zero represents a column of $n - 1$ zeros, and B_2 is $(n - 1)$-rowed square matrix. By Lemma 6 $|A| = |B| = 0$, and by Lemma 9 $|B| = a_{11}|B_2|$. Since $a_{11} \neq 0$, the number $|B_2| = 0$. We may then solve the homogeneous system of $n - 1$ equations in $n - 1$ unknowns with B_2 as matrix and obtain a nontrivial solution $(x_2, \ldots, x_n)$. Determine x_1 by formula (43), and obtain a nontrivial solution $(x_1, \ldots, x_n)$ of the original system.

13. The characteristic equation. If $A = (a_{ij})$ is any n-rowed square matrix, we may subtract x from the diagonal elements of A and thus form the matrix

$$(A - xI) = \begin{pmatrix} a_{11} - x & a_{12} & \cdots & a_{1n} \\ a_{21} & a_{22} - x & \cdots & a_{2n} \\ \cdots & \cdots & \cdots & \cdots \\ a_{n1} & a_{n2} & \cdots & a_{nn} - x \end{pmatrix}.$$

Compute the determinant of this matrix. It is a polynomial in x whose leading coefficient is $(-1)^n$. It follows that the polynomial

$$f(x) = (-1)^n|A - xI| = x^n + a_1x^{n-1} + \cdots + a_n$$

has leading coefficient unity. We call $f(x)$ the *characteristic determinant* of A. It is actually equal to $|xI - A|$.

The principal minors of A are the determinants of those square submatrices of A whose diagonal elements are all diagonal elements of A. It can be shown that the coefficient a_i of $f(x)$ is actually equal to $(-1)^i$ times the sum of all i-rowed principal minors of A. We shall not prove this result, but it is important to observe that

$$-a_1 = a_{11} + \cdots + a_{nn}$$

is the sum of the diagonal elements of A and that

$$f(0) = a_n = |-A| = (-1)^n |A|.$$

The equation $f(x) = 0$ is called the *characteristic equation* of A and its roots are called the *characteristic roots* of A.

EXERCISE

Compute the characteristic determinants and characteristic roots of the following matrices:

(a) $\begin{pmatrix} 0 & 1 & 0 \\ 0 & 0 & 1 \\ 2 & 1 & -2 \end{pmatrix}$ (b) $\begin{pmatrix} 4 & 0 & 0 \\ 0 & 0 & 1 \\ 0 & -1 & 2 \end{pmatrix}$

(c) $\begin{pmatrix} 0 & 1 & 0 & 0 \\ 0 & 0 & 1 & 0 \\ 0 & 0 & 0 & 1 \\ 1 & 4 & -6 & 4 \end{pmatrix}$ (d) $\begin{pmatrix} 0 & 0 & 0 & 4 \\ 1 & 0 & 0 & 0 \\ 0 & 1 & 0 & 5 \\ 0 & 0 & 1 & 0 \end{pmatrix}$

(e) $\begin{pmatrix} 0 & 3 & 0 \\ 3 & 3 & 3 \\ 0 & 3 & 0 \end{pmatrix}$ (f) $\begin{pmatrix} 6 & -1 & -3 \\ -1 & 0 & -3 \\ -3 & -3 & -2 \end{pmatrix}$

14. Similar matrices. Two n-rowed square matrices A and B are said to be *similar* if there exists a nonsingular matrix L such that

$$B = L^{-1}AL.$$

Theorem 10. *Similar matrices have the same characteristic determinants and consequently the same characteristic roots.*

For $|B - xI| = |L^{-1}AL - xI| = |L^{-1}(A - xI)L| = |L^{-1}| \cdot |A - xI| |L| = |A - xI|$. Then if A and B are similar, their characteristic determinants are the same, and we have proved the theorem.

We also can prove the following:

Theorem 11. *Let* A *be similar to a diagonal matrix* D. *Then the diagonal elements of* D *are the characteristic roots of* A.

For if $D = diag\{d_1, \ldots, d_n\}$, the matrix $D - xI = diag\{d_1 - x, \ldots, d_n - x\}$ and, by Lemma 8, $|D - xI| = (d_1 - x) \cdots (d_n - x)$. But $|D - xI| = |A - xI|$ and $d_1, \ldots, d_n$ must be the characteristic roots of A.

Theorem 11 states that, while it is actually not true that all square matrices are similar to diagonal matrices, if a matrix A

is similar to a diagonal matrix D the matrix D is uniquely determined up to an arbitrary permutation of its diagonal elements. Suppose then that A is given and that we determine the characteristic roots of A and so prescribe D. We propose to try to find a nonsingular matrix L such that

$$AL = LD.$$

The jth column of AL is found by our row by column rule of matrix multiplication to be the product $AQ^{(j)}$, where $Q^{(j)}$ is the jth column of L. The jth column of LD is the product $Q^{(j)}d_j$ of the jth column of L by the jth diagonal element d_j of D. Then $AL = LD$ if and only if $AQ^{(j)} = d_jQ^{(j)}$. But this latter equation is equivalent to the equation

$$(A - d_jI)Q^{(j)} = 0.$$

We are thus led to the determination of each column of L as a vector that is a solution of a linear homogeneous system with matrix $A - dI$, where d is a corresponding characteristic root. Since $|A - dI| = 0$ for each d, a corresponding nonzero vector Q can always be determined. A set of solutions $Q^{(1)}, \ldots, Q^{(n)}$, which are the columns of a nonsingular matrix, can be found only when A is similar to a diagonal matrix.

EXERCISE

Find a nonsingular matrix L such that $L^{-1}AL$ is a diagonal matrix in each of the following cases where L exists:

(a) $\begin{pmatrix} 0 & 1 & 0 \\ 0 & 0 & 1 \\ 2 & 1 & -2 \end{pmatrix}$
(b) $\begin{pmatrix} 0 & 1 & 0 \\ 0 & 0 & 1 \\ -1 & -3 & 3 \end{pmatrix}$

(c) $\begin{pmatrix} 0 & 1 & 0 & 0 \\ 0 & 0 & 1 & 0 \\ 0 & 0 & 0 & 1 \\ 4 & 0 & 5 & 0 \end{pmatrix}$
(d) $\begin{pmatrix} 0 & 1 & 0 \\ 1 & 1 & 1 \\ 0 & 1 & 0 \end{pmatrix}$

15. Real symmetric matrices. A matrix is called a *symmetric* matrix if it is equal to its transpose. Then a symmetric matrix is necessarily a square matrix $A = (a_{ij})$ such that $a_{ij} = a_{ji}$ for every i and j.

Theorem 12. *The characteristic roots of a real symmetric matrix are all real.*

For let $a + bi$ be a characteristic root of A where a and b are real and $i^2 = -1$. Then $A - (a + bi)I$ is singular and so is the product

$$B = [A - (a + bi)I][A - (a - bi)I]$$
$$= A^2 - 2aA + (a^2 + b^2)I = (A - aI)^2 + b^2I.$$

The matrix B is a singular matrix with real elements and, by Theorem 9, there exists a real nonzero vector P such that $PB = 0$. Then

$$PBP^* = P(A - aI)^2P^* + b^2PP^* = QQ^* + b^2PP^* = 0,$$

where $Q = P(A - aI)$ and $Q^* = (A - aI)^*P^* = (A - aI)P$, since A is symmetric. Since Q is a real vector, $QQ^* = Q \cdot Q \geqq 0$. But $PP^* = P \cdot P > 0$ and $QQ^* + b^2PP^* = 0$ is impossible if $b \neq 0$. Hence, $b = 0$, and we have proved that all characteristic roots $a + bi$ of A are real.

16. Orthogonal matrices. A matrix L is called an *orthogonal* matrix if L is an n-rowed square matrix such that $LL^* = I$ is the n-rowed identity matrix. Then

$$LL^* = L^*L = I, \qquad L^{-1} = L^*.$$

A set of m distinct vectors $P_1, \ldots, P_m$ is said to consist of pairwise orthogonal vectors if $P_i \cdot P_j = 0$ for $i \neq j$, that is, any two distinct vectors in the set are orthogonal. Let $P_1, \ldots, P_n$ be the rows of an orthogonal matrix L. Then the element in the ith row and jth column of LL^* is $P_i \cdot P_j$. Since $LL^* = I$, we see that

$$P_i \cdot P_i = 1, \qquad P_i \cdot P_j = 0 \quad (i \neq j; i, j = 1, \ldots, n).$$

Thus *the rows of an orthogonal matrix are pairwise orthogonal unit vectors*. Conversely if the rows of L are n pairwise orthogonal n-dimensional unit vectors, L is an orthogonal matrix.

The columns of L are the rows of L^*. Since $L^*L = I$, we see that the columns of an orthogonal matrix are also n pairwise orthogonal unit vectors. Conversely, if the columns of a square matrix L are pairwise orthogonal unit vectors, L is an orthogonal matrix.

Theorem 13. *Let* L *and* M *be* n*-rowed orthogonal matrices. Then* LM *is an orthogonal matrix.*

For $LL^* = I$, $MM^* = I$, $(LM)(LM)^* = LMM^*L^* = LIL^* = LL^* = I$.

Theorem 14. *Let* L *and* M *be orthogonal matrices. Then*

$$\begin{pmatrix} L & 0 \\ 0 & M \end{pmatrix}$$

is an orthogonal matrix.

For proof we merely compute

$$\begin{pmatrix} L & 0 \\ 0 & M \end{pmatrix}\begin{pmatrix} L & 0 \\ 0 & M \end{pmatrix}^* = \begin{pmatrix} L & 0 \\ 0 & M \end{pmatrix}\begin{pmatrix} L^* & 0 \\ 0 & M^* \end{pmatrix} = \begin{pmatrix} LL^* & 0 \\ 0 & MM^* \end{pmatrix} = I.$$

Note that in Theorem 3 the matrices L and M must have the same size but that this is not necessary in Theorem 14.

Theorem 15. *Let* $P_1, \ldots, P_m$ *be* m *pairwise orthogonal* n-*dimensional nonzero vectors. Then* $m \leqq n$ *and there exists an orthogonal matrix whose first* m *rows* (columns) *are scalar multiples of* $P_1, \ldots, P_m$.

For the equations $P_1 \cdot P = 0, \ldots, P_m \cdot P = 0$ form a homogeneous system of m linear equations in the coordinates $x_1, \ldots, x_n$ of P. If $m < n$, there exists a solution $P \neq 0$ of this system. It follows that if $m < n$ we can find a vector $P_{m+1} \neq 0$ such that $P_1, \ldots, P_{m+1}$ are pairwise orthogonal. If $m + 1 < n$, we can extend the set of vectors again and thus ultimately obtain vectors $P_{m+1}, \ldots, P_n$ such that $P_1, \ldots, P_n$ are n pairwise orthogonal nonzero vectors. Define the scalar multiples

$$U_i = \frac{1}{\sqrt{P_i \cdot P_i}} P_i \qquad (i = 1, \ldots, n).$$

Then $U_1, \ldots, U_n$ are n pairwise orthogonal unit vectors and are the rows of an orthogonal matrix L as well as the columns of the orthogonal matrix L^*. If $m \geqq n$, we form L as above and $P_i \cdot P_j = 0$ for $i = 1, \ldots, n$, and $j > n$ implies that $U_i \cdot P_j = 0$ for $i = 1, \ldots, n$ and $j > n$. But then $LP_j^* = 0$. However, L is nonsingular and necessarily $P_j^* = 0$, contrary to hypothesis. It follows that $m \leqq n$.

EXERCISE

Find an orthogonal matrix whose first two rows are scalar multiples of P_1 and P_2 in the following cases:

(a) $P_1 = (1, 2, -2), P_2 = (2, 1, 2)$

(b) $P_1 = (1, 0, 1), P_2 = (1, 0, -1)$

(c) $P_1 = (1, 2, 3), P_2 = (1, -5, 3)$

(d) $P_1 = (1, 1, 1), P_2 = (0, 1, -1)$

(e) $P_1 = (6, 3, 2), P_2 = (1, 0, -3)$

17. Orthogonal reduction of a symmetric matrix. A real symmetric matrix A is said to be *orthogonally equivalent* to a real symmetric matrix B if there exists an orthogonal matrix L with real elements such that

$$B = L^*AL.$$

Then $B = L^{-1}AL$ is similar to A and has the same characteristic roots as A.

Theorem 16. *Every real symmetric matrix* A *is orthogonally equivalent to the diagonal matrix* $D = \text{diag } \{d_1, \ldots, d_n\}$, *where* $d_1, \ldots, d_n$ *are the characteristic roots of* A *arranged in any prescribed order.*

For the equation $(A - d_1 I)P = 0$ has a nonzero solution P. Then P may be taken to be a unit vector and, by Theorem 15, we may determine an orthogonal matrix L_0 whose first column is $P = U_1$. Designate the jth column of L_0 by U_j, and see that

$$U_i^*AU_1 = d_1 U_i^*U_1.$$

Then $U_i^*AU_1 = b_{i1}$ is zero for $i \neq 1$, and $b_{11} = d_1$. It follows that the element b_{i1} in the ith row and first column of $B = L_0^*AL_0$ is zero if $i \neq 1$ and $b_{11} = d_1$. Since $B^* = L_0^*A^*L_0 = L_0^*AL_0 = B$ is a symmetric matrix, we have

$$B = \begin{pmatrix} d_1 & 0 \\ 0 & A_1 \end{pmatrix},$$

where A_1 is an $(n-1)$-rowed real symmetric matrix. Also

$$|B - xI| = \begin{vmatrix} d_1 - x & 0 \\ 0 & A_1 - xI \end{vmatrix} = (d_1 - x)|A_1 - xI| = |A - xI|,$$

and $d_2, \ldots, d_n$ are the characteristic roots of A_1. By the proof above applied to A_1 there exists an $(n-1)$-rowed orthogonal matrix L_1 such that

$$L_1^*A_1 L_1 = \begin{pmatrix} d_2 & 0 \\ 0 & A_2 \end{pmatrix},$$

where A_2 is an $(n-2)$-rowed real symmetric matrix whose characteristic roots must be $d_3, \ldots, d_n$. Then

$$N = L_0 \begin{pmatrix} 1 & 0 \\ 0 & L_1 \end{pmatrix}$$

is the product of two orthogonal matrices and is an orthogonal matrix having the property that

$$N^*AN = \begin{pmatrix} 1 & 0 \\ 0 & L_1 \end{pmatrix} (B) \begin{pmatrix} 1 & 0 \\ 0 & L_1^* \end{pmatrix} = \begin{pmatrix} d_1 & 0 & 0 \\ 0 & d_2 & 0 \\ 0 & 0 & A_2 \end{pmatrix}.$$

A repetition of this process yields the result desired.

The proof above is not intended to provide a procedure for finding an orthogonal matrix L such that $L^*AL = diag\ \{d_1 \ldots, d_n\}$. The procedure that is indicated by the proof is not recommended, because it is certainly too clumsy for use. The best procedure to employ is that of Sec. 14 where we keep in mind at every stage that the columns of the matrix L which we are determining are pairwise orthogonal nonzero vectors (see Sec. 4 of Chap. 7 for examples of this procedure).

18. Uniqueness of characteristic unit vectors. If d_j is a characteristic root of a symmetric matrix A and P_j is a solution of the equation $(A - d_jI)P_j^* = 0$, we call P_j a *characteristic vector* of A corresponding to the root d_j. We now prove the following:

Theorem 17. *Let* d_i *be a simple root of the characteristic equation of* A. *Then the corresponding characteristic unit vector is unique apart from sign.*

For we know that there exists an orthogonal matrix L such that $L^*AL = diag\ \{d_1, \ldots, d_n\}$. Then $AP_j^* = d_jIP_j^*$ is equivalent to $L^*AL\ L^*P_j^* = d_jL^*P_j^*$ and therefore to

$$(diag\ \{d_1, \ldots, d_n\})Q_j^* = d_jQ_j^*,$$

where $Q_j^* = L^*P_j^*$, $P_j^* = LQ_j^*$. But then

$$(diag\ \{d_1 - d_j, d_2 - d_j, \ldots, d_n - d_j\})Q_j^* = 0.$$

This means that if $Q_j = (\lambda_1, \ldots, \lambda_n)$, then $\lambda_ic_i = 0$, where $c_i = d_i - d_j$. Then $\lambda_i = 0$, for $i \neq j$, Q_j^* is a vector with λ_j in the jth row and zeros elsewhere, and $P_j^* = \lambda_jU_j^*$ where U_j^* is the jth column of L. If P_j is a unit vector, it follows that $P_j \cdot P_j = \lambda_j^2U_j \cdot U_j = \lambda_j^2 = 1$, $\lambda_j = \pm1$, $P_j = \pm U_j$ as desired.

We note that if d_j is a double root and thus is equal to d_k for some value of k, then Q_j is a vector with λ_j in the jth row, λ_k in the kth row and zeros elsewhere, and $P_j = \lambda_jU_j + \lambda_kU_k$. Thus P_j is not unique. Indeed we may show similarly that

$P_k = \mu_j U_j + \mu_k U_k.$ Then

$$P_j \cdot P_j = \lambda_j{}^2 + \lambda_k{}^2 = 1, \qquad Q_j \cdot Q_j = \mu_j{}^2 + \mu_k{}^2 = 1$$

since $U_j \cdot U_k = U_k \cdot U_j = 0.$ If also $P_j \cdot P_k = 0,$ we have

$$\lambda_j \mu_j + \lambda_k \mu_k = 0.$$

It follows that the matrix

$$\begin{pmatrix} \lambda_j & \mu_j \\ \lambda_k & \mu_k \end{pmatrix}$$

is an orthogonal matrix and therefore P_j and P_k are expressible in terms of U_j and U_k by means of a two-rowed orthogonal matrix in the case where d_j is a double root. We shall not consider further cases of this study but pass on to its application to three-dimensional geometry.

CHAPTER 7
ROTATIONS OF AXES AND APPLICATIONS

1. Orthogonal transformations. If four points in space are not coplanar, they determine a tetrahedron, *i.e.*, a pyramid with a triangular base. In particular, the origin and the three unit points

$$U = (1, 0, 0), \qquad V = (0, 1, 0), \qquad W = (0, 0, 1),$$

on the positive rays of the coordinate axes, determine a tetrahedron the vertex of which is the origin and which is called the *tetrahedron of reference*. Conversely, if a tetrahedron of reference is given, the coordinate system is completely determined. For the tetrahedron determines the origin, the unit distance, and the positive rays on the coordinate axes.

Let a coordinate system called the *initial* system be given so that every point P has initial coordinates x, y, z and P is the linear combination

$$(1) \qquad P = (x, y, z) = xU + yV + zW,$$

with coefficients the coordinates of P. We propose the study of all other rectangular Cartesian coordinate systems with the same origin O as the initial system. Every such transformed system is determined by three new unit vectors U', V', W' and every point P has transformed coordinates x', y', z'. Then

$$(2) \qquad P = x'U' + y'V' + z'W'$$

where the coefficients x', y', z' are the transformed coordinates of P. Let the initial coordinates of U', V', W' be given by

$$(3) \quad U' = (\lambda_1, \mu_1, \nu_1), \qquad V' = (\lambda_2, \mu_2, \nu_2), \qquad W' = (\lambda_3, \mu_3, \nu_3).$$

Then $P = x'(\lambda_1, \mu_1, \nu_1) + y'(\lambda_2, \mu_2, \nu_2) + z'(\lambda_3, \mu_3, \nu_3) = (x, y, z)$ and the relations between the initial and transformed coordinates of all points are given completely by the set of equations

$$(4) \qquad \begin{aligned} x &= \lambda_1 x' + \lambda_2 y' + \lambda_3 z' \\ y &= \mu_1 x' + \mu_2 y' + \mu_3 z' \\ z &= \nu_1 x' + \nu_2 y' + \nu_3 z'. \end{aligned}$$

103

These equations may be written in matrix form as

$$(5) \qquad \begin{pmatrix} x \\ y \\ z \end{pmatrix} = L \begin{pmatrix} x' \\ y' \\ z' \end{pmatrix} = \begin{pmatrix} \lambda_1 & \lambda_2 & \lambda_3 \\ \mu_1 & \mu_2 & \mu_3 \\ \nu_1 & \nu_2 & \nu_3 \end{pmatrix} \begin{pmatrix} x' \\ y' \\ z' \end{pmatrix}.$$

Then the columns of L are the pairwise orthogonal unit vectors U', V', W', and L is an orthogonal matrix.

A system of equations of the form given in formula (5) with L an orthogonal matrix is called an *orthogonal transformation* of coordinates. We have proved that any two rectangular coordinate systems with the same origin are related by an orthogonal transformation. Conversely, an orthogonal transformation may be interpreted as relating two coordinate systems in which the columns of the matrix L give the initial coordinates of the unit vectors on the transformed coordinate axes.

If L is orthogonal, $L^{-1} = L$. Then the solved form of formula (5) is

$$(6) \qquad \begin{pmatrix} x' \\ y' \\ z' \end{pmatrix} = L^* \begin{pmatrix} x \\ y \\ z \end{pmatrix} = \begin{pmatrix} \lambda_1 & \mu_1 & \nu_1 \\ \lambda_2 & \mu_2 & \nu_2 \\ \lambda_3 & \mu_3 & \nu_3 \end{pmatrix} \begin{pmatrix} x \\ y \\ z \end{pmatrix}.$$

These equations may be written in full as

$$(7) \qquad \begin{aligned} x' &= \lambda_1 x + \mu_1 y + \nu_1 z \\ y' &= \lambda_2 x + \mu_2 y + \nu_2 z \\ z' &= \lambda_3 x + \mu_3 y + \nu_3 z. \end{aligned}$$

In our applications of this theory to quadric surfaces we will determine vectors of integers which are scalar multiples of U', V', W'. The scalar multipliers are square roots of rational numbers and appear as common denominators of each equation in formula (7); this is not true of formula (4). Then it will be more convenient for us to express our final answers in the form given by formula (7). It should be noted that

$$\begin{aligned} U &= (1, 0, 0) \rightarrow (\lambda_1, \lambda_2, \lambda_3) \\ V &= (0, 1, 0) \rightarrow (\mu_1, \mu_2, \mu_3) \\ W &= (0, 0, 1) \rightarrow (\nu_1, \nu_2, \nu_3) \end{aligned}$$

where the set of coordinates of each of our points given after the arrow is the set of x', y', z' coordinates. These coordinates are obtained by substitution of the set of coordinates before the arrow for x, y, z in formula (7).

We have now shown that the columns of L are the x, y, z coordinates of the unit vectors on the x', y', z' axes and that the rows of L are the $x'. y', z'$ coordinates of the unit vectors on the x, y, z axes.

EXERCISES

1. Convert the matrices

$$A = \begin{pmatrix} 2 & -2 & 1 \\ 1 & 2 & 2 \\ 2 & 1 & -2 \end{pmatrix}, \quad B = \begin{pmatrix} 6 & 3 & 2 \\ -3 & 2 & 6 \\ 2 & -6 & 3 \end{pmatrix}$$

into orthogonal matrices that are scalar multiples αA, βB of the given matrices.

2. Let L in (5) be the orthogonal matrix αA of Exercise 1. Give the x', y', z' coordinates of the points whose x, y, z coordinates are

(a) $(-1, 2, 2)$ (d) $(-1, 1, 0)$
(b) $(0, 1, 1)$ (e) $(-1, 1, -2)$
(c) $(1, 1, 1)$ (f) $(6, -3, 2)$

3. Let the coordinates given in Exercise 2 be the x', y', z' coordinates of points. What are their x, y, z coordinates?

4. Give an x', y', z' equation of each of the following planes using the transformation of Exercise 2:

(a) $2x + y + 2z = 1$ (c) $2x - y = 1$
(b) $2x - 2y + z = 3$ (d) $x + y + z = 0$

5. Give the x', y', z' equations in parametric form of the lines joining the following pairs of points where the transformation of Exercise 2 is used and the coordinates given are x, y, z coordinates:

(a) $(-1, 2, 2)$, $(2, 1, 3)$ (d) $(3, 4, 1)$, $(-3, -4, -1)$
(b) $(1, 2, -2)$, $(0, 0, 0)$ (e) $(1, 0, -1)$, $(0, 1, 0)$
(c) $(2, 1, 2)$, $(-2, 2, 1)$ (f) $(2, 3, -4)$, $(1, 2, 3)$

6. Let L be the orthogonal matrix βB of Exercise 1. Give the following equations:
   (a) The x', y', z' equation of the plane $x = 0$.
   (b) The x', y', z' equation of the plane $y = -7$.
   (c) The x', y', z' equation of the plane $z = 1$.
   (d) The x, y, z equation of the plane $y' = 0$.
   (e) The x, y, z equation of the plane $z' = 7$.
   (f) The x, y, z equation of the plane $x' = -1$.
   (g) The x, y, z equations of the line $x' = y' = 1$.

2. Products of orthogonal transformations. The result of applying two successive orthogonal transformations of coordinates is an orthogonal transformation of coordinates called their *product*.   Thus the product of the orthogonal transformation

(8)
$$\begin{pmatrix} x \\ y \\ z \end{pmatrix} = L \begin{pmatrix} x' \\ y' \\ z' \end{pmatrix}$$

with matrix L and the orthogonal transformation

(9)
$$\begin{pmatrix} x' \\ y' \\ z' \end{pmatrix} = M \begin{pmatrix} x'' \\ y'' \\ z'' \end{pmatrix}$$

with matrix M is the product orthogonal transformation

(10)
$$\begin{pmatrix} x \\ y \\ z \end{pmatrix} = N \begin{pmatrix} x'' \\ y'' \\ z'' \end{pmatrix}$$

whose matrix N is obtained by substitution in formula (8) of the values of x', y', z' in terms of x'', y'', z'' as given by formula (9). But then

$$\begin{pmatrix} x \\ y \\ z \end{pmatrix} = L \left[M \begin{pmatrix} x'' \\ y'' \\ z'' \end{pmatrix} \right] = (LM) \begin{pmatrix} x'' \\ y'' \\ z'' \end{pmatrix}$$

and we have proved that $LM = N$, that is, *the matrix of a product of two orthogonal transformations is the product of the matrices of the transformations.*

EXERCISE

Give the equations of the orthogonal transformation with matrix $(\alpha A)(\beta B)$ of Exercise 1 of Sec. 1.

3. Reflections and rotations.   If the direction on a coordinate axis is changed, the resulting transformation is an orthogonal transformation of coordinates defined by one of the matrices

$$\begin{pmatrix} -1 & 0 & 0 \\ 0 & 1 & 0 \\ 0 & 0 & 1 \end{pmatrix}, \quad \begin{pmatrix} 1 & 0 & 0 \\ 0 & -1 & 0 \\ 0 & 0 & 1 \end{pmatrix}, \quad \begin{pmatrix} 1 & 0 & 0 \\ 0 & 1 & 0 \\ 0 & 0 & -1 \end{pmatrix}.$$

Such a transformation is called a *reflection* of axes.

Suppose now that the vertex O of the tetrahedron of reference is held fixed in space and that the tetrahedron is rotated about this vertex to some new position. The resulting orthogonal transformation is called a *rotation* of axes. It should be evident that every product of a finite number of rotations of axes is a rotation of axes.

A rotation of axes about the z axis may be conceived of as an ordinary rotation of axes of plane analytic geometry. Its equations are

$$(11) \qquad \begin{aligned} x &= ux' - vy' \\ y &= vx' + uy' \\ z &= z' \end{aligned}$$

where $u = \cos\theta, v = \sin\theta$, and the angle of rotation θ is measured in a counterclockwise direction from the unit point U to the unit point U'. The matrix form of this rotation is given by formula (5) where

$$(12) \qquad L = \begin{pmatrix} u & -v & 0 \\ v & u & 0 \\ 0 & 0 & 1 \end{pmatrix}$$

and the determinant of L is 1. Similarly the rotations about the x and y axes are space rotations whose matrices

$$(13) \qquad \begin{pmatrix} 1 & 0 & 0 \\ 0 & u & -v \\ 0 & v & u \end{pmatrix}, \qquad \begin{pmatrix} u & 0 & v \\ 0 & 1 & 0 \\ -v & 0 & u \end{pmatrix}$$

are orthogonal matrices having determinant 1. We shall call any rotation of axes about a coordinate axis a *planar* rotation of axes and shall discuss such rotations further in Chap. 8.

The product of two reflections of axes is a rotation of axes; for example,

$$\begin{pmatrix} 1 & 0 & 0 \\ 0 & 1 & 0 \\ 0 & 0 & -1 \end{pmatrix}\begin{pmatrix} 1 & 0 & 0 \\ 0 & -1 & 0 \\ 0 & 0 & 1 \end{pmatrix} = \begin{pmatrix} 1 & 0 & 0 \\ 0 & -1 & 0 \\ 0 & 0 & -1 \end{pmatrix}$$

is the matrix of a rotation of axes. Indeed the product of two reflections of distinct axes may be seen geometrically to be the planar rotation about the remaining axis through 180°. In the example above, the result of replacing y by $-y$ and z by $-z$ is the same as the rotation about the x axis through 180°.

A reflection of axes is not a rotation of axes. For a rotation of axes that carries two of the points U, V, W into a corresponding pair of the set U', V', W' is a rigid motion of the tetrahedron and must carry the remaining point of U, V, W into the remaining point of U', V', W'. We shall use this result in the derivation of the following basic theorem.

Theorem 1. *An orthogonal transformation is a rotation of axes if and only if the determinant of its matrix is 1. Every rotation of axes can be expressed as a product of three planar rotations and every orthogonal transformation not a rotation of axes can be expressed as the product of a rotation of axes and a reflection of axes.*

For consider an orthogonal transformation in which U, V, W are the unit vectors on the positive rays of the x, y, z axes and U', V', W' are the unit vectors on the positive rays of the x', y', z' axes. Apply a planar rotation about the z axis which carries the x axis into the line of intersection of the x', y' plane with the x, y plane. This rotation determines an x_1, y_1, z_1 coordinate system in which the z_1 axis coincides with the z axis. Rotate about the x_1 axis, which is in the x', y' plane, so that the z_1 axis is carried into the $x'y'$ plane. The result is an x_2, y_2, z_2 coordinate system in which the x_2, y_2 plane coincides with the x', y' plane. If the angle from the corresponding unit vector U_2 on the x_2 axis to the unit vector V_2 on the y_2 axis is not measured in the same direction as the angle from U' to V', we increase the planar rotation just described by 180° and replace V_2 by $-V_2$ and thus restore the angular orientation of the $x'y'$ plane. Thus we may assume that the angle from U_2 to V_2 is measured in the same direction as that from U' to V'. A rotation about the z_2 axis that carries U_2 into U' also carries V_2 into V'.

We have now proved the existence of three planar rotations with corresponding matrices L_1, L_2, L_3 such that the product of these rotations is a rotation of axes

$$(14) \qquad \begin{pmatrix} x \\ y \\ z \end{pmatrix} = L_1 L_2 L_3 \begin{pmatrix} x'' \\ y'' \\ z'' \end{pmatrix} = N \begin{pmatrix} x'' \\ y'' \\ z'' \end{pmatrix}$$

with matrix $N = L_1 L_2 L_3$ and $U'' = U'$, $V'' = V'$. Then W'' is a point on a line through 0 perpendicular to the x', y' plane and $W' = \epsilon W''$, where

$$(15) \qquad\qquad \epsilon = \pm 1.$$

The transformation defined by

(16)
$$\begin{pmatrix} x' \\ y' \\ z' \end{pmatrix} = L * \begin{pmatrix} x \\ y \\ z \end{pmatrix} = L * N \begin{pmatrix} x'' \\ y'' \\ z'' \end{pmatrix}$$

is an orthogonal transformation such that $x' = x''$, $y' = y''$, $z' = \epsilon z''$. Then

(17)
$$R = L * N = \begin{pmatrix} 1 & 0 & 0 \\ 0 & 1 & 0 \\ 0 & 0 & \epsilon \end{pmatrix} = N * L = N^{-1}L,$$

and

(18)
$$L = NR = L_1 L_2 L_3 \begin{pmatrix} 1 & 0 & 0 \\ 0 & 1 & 0 \\ 0 & 0 & \epsilon \end{pmatrix}.$$

Since $|N| = 1$, we see that $|L| = |R|$,

(19)
$$\epsilon = |L|.$$

The matrix R is the matrix of a rotation of axes only when $\epsilon = 1$, $R = I$. Since N is the matrix of a rotation of axes and $R = L*N$, we see that R is the matrix of a rotation of axes if and only if L is the matrix of a rotation of axes. Then L is the matrix of a rotation of axes if and only if $\epsilon = |L| = 1$. When $|L| = -1$, we see that $L = NR$ is the product of the matrix of a rotation of axes and the matrix R of a reflection of axes. This proves the theorem.

EXERCISES

Determine whether or not the following matrices are the matrices of rotations of axes:

(a) $\dfrac{1}{3} \begin{pmatrix} -2 & 1 & 2 \\ 2 & 2 & 1 \\ 1 & -2 & 2 \end{pmatrix}$

(b) $\dfrac{1}{7} \begin{pmatrix} 2 & 3 & 6 \\ 6 & 2 & -3 \\ 3 & -6 & 2 \end{pmatrix}$

(c) $\dfrac{1}{3} \begin{pmatrix} -2 & 2 & 1 \\ 2 & 1 & 2 \\ 1 & 2 & -2 \end{pmatrix}$

(d) $\begin{pmatrix} \dfrac{1}{\sqrt{3}} & \dfrac{-1}{\sqrt{3}} & \dfrac{1}{\sqrt{3}} \\ \dfrac{1}{\sqrt{2}} & \dfrac{1}{\sqrt{2}} & 0 \\ \dfrac{-1}{\sqrt{6}} & \dfrac{1}{\sqrt{6}} & \dfrac{2}{\sqrt{6}} \end{pmatrix}$

4. Orthogonal reduction of a real quadratic form. A real quadratic form in x, y, z is a polynomial

$$(20) \qquad f(x, y, z) = ax^2 + by^2 + cz^2 + 2dxy + 2exz + 2gyz,$$

where a, b, c, d, e, g are real numbers. Then

$$(21) \qquad\qquad f(x, y, z) = PAP^*$$

where $P = (x, y, z)$ and A is the real symmetric matrix

$$(22) \qquad \begin{pmatrix} a & d & e \\ d & b & g \\ e & g & c \end{pmatrix}.$$

A rotation of axes is a linear transformation

$$(23) \qquad\qquad P^* = LQ^*$$

where L is an orthogonal matrix of determinant 1 and $Q = (x', y', z')$. Then

$$(24) \qquad\qquad P = QL^*$$

and a rotation of axes replaces $f(x, y, z)$ by

$$(25) \quad \phi(x', y', z') = a'x'^2 + b'y'^2 + c'z'^2 + 2d'x'y' \\ + 2e'x'z' + 2g'y'z'$$

where

$$(26) \qquad g(x', y', z') = f(x, y, z) = QL^*ALQ^* = QBQ^*.$$

Hence

$$(27) \qquad\qquad B = \begin{pmatrix} a' & d' & e' \\ d' & b' & g' \\ e' & g' & c' \end{pmatrix} = L^*AL.$$

Theorem 2. *Every real quadratic form* PAP^* *may be reduced to a real diagonal form*

$$(28) \qquad\qquad \alpha x'^2 + \beta y'^2 + \gamma z'^2$$

by a rotation of axes where α, β, γ *are the characteristic roots of* A.

For, by Theorem 16 of Chap. 6, there exists an orthogonal matrix L such that

$$L_0^*AL_0 = \begin{pmatrix} \alpha & 0 & 0 \\ 0 & \beta & 0 \\ 0 & 0 & \gamma \end{pmatrix}.$$

If $|L_0| = 1$, we take $L = L_0$ and the orthogonal transformation with matrix L is a rotation of axes. If $|L_0| = -1$, we may change the sign of one column of L_0 and replace L_0 by an orthogonal matrix L such that $|L| = -|L_0| = 1$. Evidently $L^*AL = L_0^*AL_0$. Then the rotation of axes with matrix L replaces $f(x, y, z)$ by $\phi(x', y', z') = Q(L^*AL)Q^* = \alpha x'^2 + \beta y'^2 + \gamma z'^2$.

ILLUSTRATIVE EXAMPLES

I. Reduce the quadratic form $4x^2 + 3y^2 - z^2 - 12xy + 4xz - 8yz$ to diagonal form $d_1 x'^2 + d_2 y'^2 + d_3 z'^2$ by a rotation of axes such that $d_1 \geqq d_2 \geqq d_3$. Give the diagonal form and the equations of rotation.

Solution

The matrix of the given form is

$$A = \begin{pmatrix} 4 & -6 & 2 \\ -6 & 3 & -4 \\ 2 & -4 & -1 \end{pmatrix},$$

and

$$|A - xI| = \begin{vmatrix} 4 - x & -6 & 2 \\ -6 & 3 - x & -4 \\ 2 & -4 & -1 - x \end{vmatrix}$$

$$= - \begin{vmatrix} 4 - x & 6 & 2 \\ 2(1 - x) & 9 + x & 0 \\ 2 & 4 & -1 - x \end{vmatrix} = -f(x).$$

Then $f(x) = 2[8(1 - x) - 2(x + 9)] - (1 + x)[(4 - x)(x + 9) - 12(1 - x)] = -20(x + 1) - (x + 1) - x^2 + (4 - 9 + 12)x + 36 - 12 = (x + 1)(x^2 - 7x - 44) = (x + 1)(x + 4)(x - 11)$. This yields the diagonal form $11x'^2 - y'^2 - 4z'^2$.

We now solve the system of equations

$$\begin{aligned} -7x - 6y + 2z &= 0 \\ -6x - 8y - 4z &= 0 \\ 2x - 4y - 12z &= 0 \end{aligned}$$

with matrix $A - (11)I$. Then $6x - 12y - 36z = 0$, $-20y - 40z = 0$, $y = -2z$, $2x = 4y + 12z = -8z + 12z = 4z$, $x = 2z$. The remaining equation merely yields the check $-14z + 12z + 2z = 0$. We have shown that $(2, -2, 1)$ is a characteristic vector of $A - (11)I$.

We next determine a characteristic vector of $A + I$ and thus solve

$$\begin{aligned} 5x - 6y + 2z &= 0 \\ -6x + 4y - 4z &= 0 \\ 2x - 4y &= 0. \end{aligned}$$

Then $x = 2y$, $-12y + 4y = 4z = -8y$, $z = -2y$. This yields the vector $(2, 1, -2)$. The final vector is a characteristic vector of $A + 4I$ but may be obtained as a vector orthogonal to the vectors already determined. Then we have $2x - 2y + z = 0$, $2x + y - 2z = 0$, $3y = 3z$, $y = z$, $2x = 2y - z = z$, and we have shown that the columns of the matrix

$$\begin{pmatrix} 2 & 2 & 1 \\ -2 & 1 & 2 \\ 1 & -2 & 2 \end{pmatrix}$$

are the required characteristic vectors. We compute

$$\begin{vmatrix} 2 & 2 & 1 \\ -2 & 1 & 2 \\ 1 & -2 & 2 \end{vmatrix} = \begin{vmatrix} 0 & 0 & 1 \\ -6 & -3 & 2 \\ -3 & -6 & 2 \end{vmatrix} = 36 - 9 = 27$$

and see that

$$L = \tfrac{1}{3}\begin{pmatrix} 2 & 2 & 1 \\ -2 & 1 & 2 \\ 1 & -2 & 2 \end{pmatrix}$$

is the matrix of a rotation of axes. The equations of this rotation in solved form are $(x', y', z') = (x, y, z)L$,

$$\begin{aligned} 3x' &= 2x - 2y + z \\ 3y' &= 2x + y - 2z \\ 3z' &= x + 2y + 2z. \end{aligned}$$

II. Reduce the quadratic form $-x^2 - y^2 - 7z^2 + 16xy + 8xz + 8yz$ to diagonal form by a rotation of axes.

Solution

The matrix of the form is

$$\begin{pmatrix} -1 & 8 & 4 \\ 8 & -1 & 4 \\ 4 & 4 & -7 \end{pmatrix}$$

and the characteristic determinant is

$$\begin{vmatrix} -1-x & 8 & 4 \\ 8 & -1-x & 4 \\ 4 & 4 & -7-x \end{vmatrix} = \begin{vmatrix} -1-x & 0 & 4 \\ 8 & -9-x & 4 \\ 4 & 2x+18 & -7-x \end{vmatrix}$$

$$= (x+9)\begin{vmatrix} -1-x & 0 & 4 \\ 8 & -1 & 4 \\ 4 & 2 & -7-x \end{vmatrix}$$

Then $f(x) = -|A - xI| = -(x + 9)[4(16 + 4) - (1 + x). (x + 7 - 8)] = (x + 9)(x^2 - 1 - 80) = (x + 9)(x + 9)(x - 9)$. The diagonal form is $9(x'^2 - y'^2 - z'^2)$.

We first solve the system

$$-10x + 8y + 4z = 0$$
$$8x - 10y + 4z = 0$$
$$4x + 4y - 16z = 0$$

with matrix $A - 9I$ and obtain $18y = 36z$, $y = 2z$, $x = 4z - y = 2z$. This yields a characteristic vector $(2, 2, 1)$. However, the system with matrix $A + 9I$ consists of the three equations

$$8x + 8y + 4z = 0$$
$$8x + 8y + 4z = 0$$
$$4x + 4y + 2z = 0.$$

Moreover, the condition that this vector be orthogonal to $(2, 2, 1)$ is $2x + 2y + z = 0$. Hence, any choice such as $(-1, 1, 0)$ is a characteristic vector. The remaining characteristic vector satisfies $2x + 2y + z = 0$ and $x - y = 0$, so that $x = y$, $z = -4y$ and $(1, 1, -4)$ is a characteristic vector,

$$L = \begin{pmatrix} \dfrac{2}{3} & \dfrac{-1}{\sqrt{2}} & \dfrac{1}{\sqrt{18}} \\[2mm] \dfrac{2}{3} & \dfrac{1}{\sqrt{2}} & \dfrac{1}{\sqrt{18}} \\[2mm] \dfrac{1}{3} & 0 & \dfrac{-4}{\sqrt{18}} \end{pmatrix}$$

is the matrix of an orthogonal transformation carrying the given quadratic form into the diagonal form above. But

$$\begin{vmatrix} 2 & -1 & 1 \\ 2 & 1 & 1 \\ 1 & 0 & -4 \end{vmatrix} = \begin{vmatrix} 0 & -2 & 0 \\ 2 & 1 & 1 \\ 1 & 0 & -4 \end{vmatrix} = 2(-8 - 1) = -18$$

and thus L is not the matrix of a rotation of axes. We therefore change the sign of the second column and obtain the following equations of rotation:

$$3x' = 2x + 2y + z$$
$$\sqrt{2}y' = x - y$$
$$3\sqrt{2}z' = x + y - 4z.$$

EXERCISES

1. Show that $(2, -1, -2)$ is a characteristic vector of the matrix of Illustrative Example II corresponding to the root -9, and obtain the corresponding equations of rotation.

2. Reduce each of the following quadratic forms to diagonal form $d_1x'^2 + d_2y'^2 + d_3z'^2$ with $d_1 = d_2 = d_3$ by a rotation of axes. Give the diagonal form and the equations of rotation solved for x', y', z' as your answer.

(a) $2x^2 + y^2 - 4xy - 4yz$

Ans. $x'^2 - 2y'^2 + 4z'^2$; $3x' = 2x + y - 2z$; $3y' = x + 2y + 2z$; $3z' = 2x - 2y + z$.

(b) $3x^2 - 3y^2 - 5z^2 - 2xy - 6xz - 6yz$

(c) $3x^2 + y^2 + z^2 - 2xy + 2xz - 2yz$

Ans. $4x'^2 + z'^2$; $\sqrt{6}x' = 2x - y + z$; $\sqrt{2}y' = y + z$; $\sqrt{3}z' = -x - y + z$.

(d) $-2x^2 + 4y^2 + 6z^2 + 2xy + 6xz + 6yz$

(e) $4x^2 + y^2 - 8z^2 + 4xy - 4xz + 8yz$

Ans. $5x'^2 + 2y'^2 - 10z'^2$; $\sqrt{5}x' = 2x + y$; $\sqrt{6}y' = -x + 2y + z$; $\sqrt{30}z' = x - 2y + 5z$.

(f) $3x^2 + 3z^2 + 4xy + 8xz + 4yz$

5. Quadric surfaces. Suppose that p, q, r, s are real numbers and that $f_0(x, y, z)$ is a real quadratic form. Then the equation

$$(29) \quad f(x, y, z) \equiv f_0(x, y, z) + 2(px + qy + rz) + s = 0$$

is the general form of an equation defining a quadric surface. By Theorem 2 there exists a rotation of axes which replaces this equation by

$$(30) \quad \phi(x', y', z') = \alpha x'^2 + \beta y'^2 + \gamma z'^2 + 2(\rho x' + \sigma y' + \tau z') + \delta = 0$$

for real numbers α, β, γ, ρ, σ, τ, δ such that α, β, γ are not all zero. The vector (α, β, γ) of the characteristic roots of the matrix of $f_0(x, y, z)$ is unique up to a nonzero real factor t that enters when we multiply $f(x, y, z) = 0$ by t. Thus we may always take

$$(31) \quad \alpha > 0, \quad \beta \geq 0, \quad \alpha \geq \beta \geq \gamma.$$

In Chap. 5 we studied all the quadric surfaces defined by the equation of formula (30) except that where $\beta = \gamma = 0$ and $\rho \neq 0$, $\tau \neq 0$. Then

$$\phi(x', y', z') = \alpha\left(x' + \frac{\rho}{\alpha}\right)^2 + 2(\sigma y' + \tau z') + \delta - \frac{\rho^2}{\alpha} = 0$$

and we may perform a second rotation of axes defined by

$$x'' = x'$$

(32)
$$y'' = \frac{\sigma y' + \tau z'}{\sqrt{\sigma^2 + \tau^2}}$$

$$z'' = \frac{-\tau y' + \sigma z'}{\sqrt{\sigma^2 + \tau^2}}.$$

This planar rotation of axes replaces $\phi(x', y', z') = 0$ by

(33) $$F(x'', y'', z'') = \alpha \left(x'' + \frac{\rho}{\alpha} \right)^2 + 2\sqrt{\sigma^2 + \tau^2}\, y''$$

$$+ \delta - \frac{\rho^2}{\alpha} = 0.$$

The translation of axes whose equations are

(34) $$x''' = x'' + \frac{\rho}{\alpha}, \qquad y''' = y'' + \frac{\alpha\delta - \rho^2}{\alpha\sqrt{\sigma^2 + \tau^2}}, \qquad z''' = z''$$

replaces $F(x'', y'', z'') = 0$ by

$$\Phi(x''', y''', z''') \equiv \alpha x'''^2 + 2\sqrt{\sigma^2 + \tau^2}\, y''' = 0.$$

This is an equation of a parabolic cylinder with vertex at the origin. Then we have proved that every quadric surface is one of the surfaces discussed in Chap. 5.

Let us close this section with a summary of the allowable operations on a quadratic equation $f(x, y, z) = 0$ defining a quadric surface, and of their effect on the quadratic form $f_0(x, y, z)$.

The first operation is that of a translation of axes. It should be evident that only the linear and constant terms of $f(x, y, z)$ are affected by this operation and that $f_0(x, y, z)$ is unaltered.

The second operation is that of a rotation of axes. This replaces $f_0(x, y, z)$ by an orthogonally equivalent quadratic form having the same characteristic roots. If A is the matrix of $f_0(x, y, z)$ and L^*AL is the matrix of the equivalent form, then $P(A - dI) = 0$ if and only if $PLL^*(A - dI)L = 0$, $(PL)(L^*AL - dI) = 0$. Thus the vector P is a characteristic vector of A corresponding to a root d if and only if PL is a characteristic vector of L^*AL corresponding to the same root.

The final operation is that of multiplying $f(x, y, z)$ by a nonzero constant k. This multiplies $f_0(x, y, z)$ by k and the matrix A of $f_0(x, y, z)$ by k. The corresponding characteristic roots are multiplied by k and indeed if $L^*AL = diag\{\alpha, \beta, \gamma\}$ then $L^*(kA)L$

$= diag\{k\alpha, k\beta, k\gamma\}$. Also $P(A - dI) = 0$ if and only if $P(kA - kdI) = 0$ and therefore the characteristic vectors corresponding to the given quadratic form are unaltered by this change.

EXERCISE

Reduce the following equations to simplified form by a rotation of axes. Give the center or vertex of the quadric, the type of quadric, the semiaxes, and the equations of rotation.

(a) $2x^2 + 2y^2 - z^2 + 8xy - 4xz - 4yz = 2$

(b) $2x^2 + y^2 + 2z^2 + 2xy - 2yz = 1$

(c) $4x^2 + 6y^2 + 4z^2 - 4xz + 1 = 0$

(d) $2x^2 + y^2 - 4xy - 4yz + 12x + 6y + 6z = 1$

(e) $x^2 + y^2 + z^2 - 4yz - 4xz - 4xy = 7$

(f) $5x^2 + 5y^2 + 3z^2 - 2xy + 2xz + 2yz + 2x - y = 0$

(g) $y^2 + z^2 - xy + xz + yz - 2x + 2y - 2z + 1 = 0$

(h) $x^2 + 4y^2 + 9z^2 - 4xy + 6xz - 12yz + 4x - 8y + 12z + 4 = 0$

(i) $16x^2 + 9y^2 + 4z^2 + 24xy - 16xz - 12yz + 7x + 2y - 12z = 0$

(j) $2x^2 + 3y + 4z + 4 = 0$

(k) $2z^2 + 5x + 12y + 12z + 18 = 0$

(l) $x^2 + y^2 + 4z^2 - 2xy - 4xz + 4yz + 6x + 12y + 18z = 0$

(m) $2x^2 + 2y^2 - 4z^2 - 5xy - 2xz - 2yz - 2x - 2y + z = 0$

(n) $3x^2 + y^2 + z^2 + 4yz + 12x + 2y - 2z + 9 = 0$

(o) $4x^2 + 4y^2 + 9z^2 + 8xy + 12xz + 12yz + 10x + y + 4z + 1 = 0$

(p) $2x^2 + 4yz + 6z + 2y - 4x + 5 = 0$

(q) $3x^2 + 3y^2 + z^2 - 2xy + 6x - 2y - 2z + 3 = 0$

6. Plane sections of quadrics. If $f(x, y, z) = 0$ is an equation of a quadric surface and $ax + by + cz + d = 0$ is an equation of a plane, we may rotate axes so that the x', y' plane is parallel to the given plane. We may then translate axes so that the given plane becomes the plane $z' = 0$. It follows that the plane sections of any quadric are the sections by the plane $z = 0$ of the general quadric of formula (29) for a properly selected coordinate system.

As a consequence every plane section of a quadric is a conic

$$(35) \quad z = 0, \quad ax^2 + by^2 + 2dxy + 2hx + 2py + r = 0,$$

as studied in plane analytic geometry. Such a conic may be carried by a rotation of axes leaving z unaltered into a conic

$$(36) \quad z = z' = 0, \quad \alpha x'^2 + \beta y'^2 + 2\gamma x' + 2\delta y' + \epsilon = 0.$$

Apply this rotation to the original quadric and thus carry the quadratic form $f_0(x, y, z) = ax^2 + by^2 + cz^2 + 2dxy + 2exz + 2gyz$ into

(37) $\phi_0(x', y', z') \equiv \alpha x'^2 + \beta y'^2 + \gamma z'^2 + 2\lambda x'z' + 2\mu y'z'.$

If the given quadric is an ellipsoid, then $\phi_0(x', y', z') > 0$ for all real values of x', y', z' not all zero and this must be true of $\phi(x', y', 0)$. Hence, formula (37) is our equation of an ellipse and we have proved the following:

Theorem 3. *The plane sections of a quadric are conic sections. In particular, all plane sections of an ellipsoid are ellipses.*

7. Points of symmetry. We shall close this chapter by applying notations of axes to obtain a discussion of the symmetries of nondegenerate quadrics, *i.e.*, quadrics that are not planes, pairs of distinct planes, point loci, or imaginary loci. *It is recommended that the details of the proofs be omitted in the classroom and that only the results be presented.*

We first observe that the nondegenerate quadrics defined by equations of the form

$$\alpha x^2 + \beta y^2 + \gamma z^2 + \delta = 0$$

are symmetric with respect to the origin. We shall call such quadrics *central quadrics* and shall say that the origin is a *center*. If $\alpha\beta\gamma \neq 0$, the corresponding surfaces are ellipsoids, hyperboloids, or quadric cones. Then the origin is the only point of symmetry. Indeed, we shall prove the following result:

Theorem 4. *A nondegenerate quadric* S *is symmetric with respect to a point* P *if and only if* S *is a central quadric with* P *as a center. If* S *is an ellipsoid, hyperboloid, or quadric cone, the point* P *is the only point of symmetry. However, if* S *is a cylinder, the line through* P *parallel to a generating line of* S *is a line of centers of* S.

For we select a coordinate system with a given point of symmetry P as origin. Then if $f(x, y, z) = 0$ is an equation of S relative to this coordinate system, it must be true that $f(x, y, z) \equiv f(-x, -y, -z)$. Hence $f(x, y, z)$ has no linear terms. A rotation of axes about P reduces $f(x, y, z)$ to the form $\alpha x^2 + \beta y^2 + \gamma z^2 + \delta = 0$ and S is a central quadric, P is a center of S. Let (ξ, η, ζ) be a second point of symmetry, and carry out a translation of axes $x = x' + \xi, y = y' + \eta, z = z' + \zeta$, which moves the origin to (ξ, η, ζ). Then $f(x, y, z) = \alpha(x' + \xi)^2 +$

$\beta(y' + \eta)^2 + \gamma(z' + \zeta)^2 = \phi(x', y', z')$, $\phi(-x', y', z') = \phi(x, y, z)$, and $\alpha\xi = \beta\eta = \gamma\zeta = 0$. If $\alpha\beta\gamma \neq 0$, then $(\xi, \eta, \zeta) = (0, 0, 0)$ and P is the only point of symmetry. Since S is nondegenerate, only one of the numbers α, β, γ can be zero, and we can assume the coordinate system chosen so that $\alpha\beta \neq 0$, $\gamma = 0$. Then $\xi = \eta = 0$ and the surface S is symmetric with respect to all points $(0, 0, \zeta)$, that is, all points on the z axis. But the z axis is the line through P parallel to the generating line $x = y = 0$ of the cylinder $\alpha x^2 + \beta y^2 + \delta = 0$. This proves the theorem.

8. Planes of symmetry. Define a quadric surface S by a quadratic equation $f(x, y, z) = 0$, and define a plane H by an equation $\lambda x + \mu y + \nu z = p$, where $p \geq 0$ and $\lambda^2 + \mu^2 + \nu^2 = 1$. Then a line L normal to H and through a point (x_0, y_0, z_0) of H is a line $x = x_0 + \lambda t$, $y = y_0 + \mu t$, $z = z_0 + \nu t$. We find the points of intersection of L and S by finding the roots of the equation

(38) $\phi(t) \equiv f(x_0 + \lambda t, y_0 + \mu t, z_0 + \nu t) = 0$.

The distance from H to any point of L is $\lambda(x_0 + \lambda t) + \mu(y_0 + \mu t) + \nu(z_0 + \nu t) - p = (\lambda^2 + \mu^2 + \nu^2)t + (\lambda x_0 + \mu y_0 + \nu z_0 - p) = t$ and thus the solutions of $\phi(t) = 0$ define not only the points of intersection of the normal lines with S but also their distances from H.

A plane H is a plane of symmetry of a quadric surface S if every normal to H either does not cut S or cuts S in two points on opposite sides of H and at the same distance from H. While this definition formally includes a plane having the property that all normals either do not cut the surface or lie wholly on it, we shall exclude such planes. Thus the planes $z = k$ are not regarded as planes of symmetry of the cylinders $f(x, y) = 0$. A plane of symmetry then defines a set of normal lines and corresponding pairs of points of intersection P_1 and P_2 whose corresponding distances t_1 and t_2 from H must have the property $t_1 + t_2 = 0$. Then necessarily the coefficient of t in $\phi(t)$ is zero for all (x_0, y_0, z_0) in H.

If f_x, f_y, f_z are the partial derivatives of $f(x, y, z)$ with respect to x, y, z, the coefficient of t in formula (38) is

$$2\phi_1(x_0, y_0, z_0) \equiv \lambda f_x(x_0, y_0, z_0) + \mu f_y(x_0, y_0, z_0) + \nu f_z(x_0, y_0, z_0).$$

We thus seek to determine all values of p, λ, μ, ν such that

$\phi_1(x_0, y_0, z_0) = 0$ for (x_0, y_0, z_0) on the plane H defined by $\lambda x + \mu y + \nu z - p = 0$.

The central quadric

$$(39) \qquad \alpha x^2 + \beta y^2 + \gamma z^2 + \delta = 0,$$

defined for real α, β, γ, δ such that $\alpha\beta\gamma \neq 0$, has the corresponding function

$$(40) \qquad \phi_1(x_0, y_0, z_0) \equiv \alpha\lambda x_0 + \beta\mu y_0 + \gamma\nu z_0.$$

If $\lambda \neq 0$, then $(p\lambda^{-1}, 0, 0)$ is in H and therefore $\alpha\lambda x_0 - \alpha p = 0$. Then $p = 0$. Similarly $p = 0$ if $\mu \neq 0$ or $\nu \neq 0$. Since $\lambda^2 + \mu^2 + \nu^2 = 1$, we have proved that H is a plane through the center of S.

If $\lambda = \mu = 0$, then $\nu \neq 0$ and H is the plane $z = 0$. If $\lambda = 0$ and $\mu \neq 0$, then $y_0 = -\mu^{-1}\nu z_0$ and $(\gamma - \beta)\nu z_0 \equiv 0$ for all values of x_0 and z_0. Then $(\gamma - \beta)\nu = 0$. If S is a sphere, any plane through the center of S is a plane of symmetry. If S is not a sphere, we may take $\gamma \neq \beta$ and see that $\nu = 0$; H is the plane $y = 0$.

There remains the case $\lambda \neq 0$, $x_0 = -\lambda^{-1}(\mu y_0 + \nu z_0)(\beta - \alpha)$ $\mu y_0 + (\gamma - \alpha)\gamma z_0 = 0$ for all y_0 and z_0. Then $(\beta - \alpha)\mu = (\gamma - \alpha)\nu = 0$. When S is not a surface of revolution, $\beta \neq \alpha$, $\gamma \neq \alpha$ so that $\mu = \nu = 0$ and H is the plane $x = 0$. If S is a surface of revolution defined by $\alpha = \beta \neq \gamma$, we have $\nu = 0$; H is a plane $\lambda x + \mu y = 0$ through the axis of revolution $x = y = 0$. But all such planes are planes of symmetry. We have proved the following:

Theorem 5. *A plane is a plane of symmetry of a sphere if and only if it is a plane through the center of the sphere. If S is a central quadric of revolution, its planes of symmetry are the planes through the axis of revolution and the plane through the center perpendicular to the axis of revolution. The only planes of symmetry of a central quadric* $\alpha x^2 + \beta y^2 + \gamma z^2 + \delta = 0$, *which is not a surface of revolution, are the planes* x = 0, y = 0, *and* z = 0.

The equation

$$(41) \qquad \alpha x^2 + \beta y^2 = 2\delta z$$

defines a paraboloid for all nonzero real numbers α, β, δ. Then

$$(42) \qquad \phi_1(x_0, y_0, z_0) \equiv \alpha\lambda x_0 + \beta\mu y_0 - \delta\nu.$$

If $\gamma \neq 0$, then $(0, 0, p\nu^{-1})$ is on H and $\phi_1(0, 0, p\nu^{-1}) = -\delta\nu = 0$.

But $\delta \neq 0$, $\nu \neq 0$, which is a contradiction. Hence, $\mu = 0$ and $\lambda x + \mu y = p$ is an equation of H, $\phi_1 \equiv \alpha \lambda x_0 + \beta \mu y_0$. If $\lambda \neq 0$, then $(\lambda^{-1}p, 0, 0)$ is on H and $\phi_1(\lambda^{-1}p, 0, 0) = \alpha p = 0$, $p = 0$. If $\lambda = 0$, then $\mu \neq 0$ and $(0, \mu^{-1}p, 0)$ is on H, $\phi_1(0, \mu^{-1}p, 0)$ $= \beta p = 0$, and $p = 0$. We have thus shown that every plane of symmetry of S is a plane $\lambda x + \mu y = 0$. If $\alpha = \beta$, so that S is a surface of revolution, all such planes are planes of symmetry. But if $\alpha \neq \beta$, we have $\lambda x_0 = -\mu y_0$, $\phi_1 = (\beta - \alpha)\mu y_0$. When $\lambda \neq 0$, $\phi_1 = (\beta - \alpha)\mu y_0 = 0$ for all values of y_0 and $\mu = 0$, H is the plane $x = 0$. Otherwise, $\lambda = 0$ and H is the plane $y = 0$. We have proved the following:

Theorem 6. *A paraboloid $\alpha x^2 + \beta y^2 = 2\delta z$ has the planes x = 0, y = 0 as planes of symmetry. These are its only planes of symmetry unless the paraboloid is a surface of revolution. In this latter case, the planes of symmetry are all the planes through the axis of revolution.*

The only remaining nondegenerate quadrics are the cylinders. The first case to be studied is that of a cylinder

(43) $\alpha x^2 + \beta y^2 = \delta$

where $\alpha\beta\delta \neq 0$. We have already seen that a plane $z = k$ is not a plane of symmetry and therefore all planes of symmetry are planes $\lambda x + \mu y + \nu z = p$, where $\lambda \neq 0$ or $\mu \neq 0$. Now $\phi_1(x_0, y_0, z_0) = \alpha \lambda x_0 + \beta \mu y_0$ and $(\lambda^{-1}p, 0, 0)$ or $(0, \mu^{-1}p, 0)$ is on H, $\phi_1(x_0, y_0, z_0) = \alpha p$ or βp at these points and therefore $p = 0$. If $\lambda = 0$, then $\mu \neq 0$, $y_0 = -\mu^{-1}\nu z_0$, $\alpha \lambda x_0 - \beta \nu z_0 = 0$ for all values of $x_0, z_0, \lambda = \nu = 0$ and H is the plane $y = 0$. If $\lambda \neq 0$, then $x_0 = -\lambda^{-1}(\mu y_0 + \nu z_0)$, $\phi_1(x_0, y_0, z_0) \equiv (\beta - \alpha)\mu y_0$ $- \alpha\nu z_0 = 0$ for all y_0 and z_0, $(\beta - \alpha)\mu = \nu = 0$. When $\beta = \alpha$, the surface S is a circular cylinder and all planes through its axis $x = y = 0$ are planes of symmetry. In cases $\beta \neq \alpha$, then $\mu = \nu = 0$ and H is the plane $x = 0$. We have proved the following result:

Theorem 7. *The planes of symmetry of a circular cylinder are the planes through its axis. If a noncomposite cylinder $\alpha x^2 + \beta y^2 = \delta$ is not a circular cylinder, the only planes of symmetry are the planes x = 0 and y = 0.*

The only remaining case is that of the parabolic cylinder $x^2 = \delta y$, where $\delta \neq 0$. As before, every plane of symmetry is a plane $\lambda x + \mu y + \nu z = p$, where $\lambda \neq 0$ or $\mu \neq 0$. Now $\phi_1(x_0,$

$y_0, z_0) = \lambda x_0 - \delta\mu$. If $\mu \neq 0$, then $(0, \mu^{-1}p, 0)$ is on H, $-\delta\mu = 0$ contrary to hypothesis. Hence, $\mu = 0$, $\lambda \neq 0$. Then $(\lambda^{-1}p, 0, 0)$ is on H, $\phi_1 = \lambda x_0 = p = 0$. It follows that H is a plane $\lambda x = -\mu z$, $(-\lambda^{-1}\nu z_0, y_0, z_0)$ is on H for all y_0, z_0, $\phi_1 = 0$, and H is the plane $x = 0$. We have proved the following:

Theorem 8. *The only plane of symmetry of a parabolic cylinder* $x^2 = \delta y$ *is the plane* x $= 0$.

9. Lines of symmetry. If L is a line of symmetry of a quadric surface S, we may select a coordinate system such that L is the z axis. Let an equation of S relative to this coordinate system be $f(x, y, z) = 0$. Then $f(x, y, z) \equiv f(-x, y, z) \equiv f(x, -y, z)$, so that $f(x, y, z)$ involves only even powers of x and y. Then

$$(44) \qquad f(x, y, z) \equiv \alpha x^2 + \beta y^2 + \gamma z^2 + 2\delta z + \epsilon = 0.$$

Suppose first that S is a central quadric with three nonzero characteristic roots. Then $\alpha\beta\gamma \neq 0$, and we may select the coordinate system so that $\delta = 0$. The given line is then a line of intersection of two distinct planes of symmetry, *i.e.*, the planes 0. When α, β, γ are distinct, the surface S has ...es of symmetry and thus the only possible lines ...re the three intersections of pairs of such planes. ...called the *principal axes* of the quadric. When ...ll lines through the center are lines of symmetry. ...s a quadric of revolution, the intersection of two ...hrough the axis of revolution is the axis of revolu- ...ersection of a plane through the axis of revolution ... plane perpendicular to the axis of revolution is a line through the center perpendicular to the axis of revolution. These lines are all lines of symmetry. We have proved the following:

Theorem 9. *The lines of symmetry of a sphere are all the lines through its center. If* S *is a noncylindrical central quadric of revolution, its lines of symmetry are the axis of revolution and all lines through the center of* S *and perpendicular to the axis of revolution. The lines of symmetry of a noncylindrical central quadric, not a surface of revolution, are its three principal axes.*

We next assume that S is a paraboloid. Then necessarily $\gamma = 0$, $\alpha \neq 0$, $\beta \neq 0$ in formula (44), and the line L of symmetry is the intersection of two planes of symmetry of the surface. By Theorem 6, if S is not a surface of revolution, L is the unique line of intersection of its two planes of symmetry. If S is a surface of revolution, all planes of symmetry intersect on the axis of revolution and L is this axis.

Theorem 10. *A paraboloid has a single line of symmetry.*

If S is a cylinder, then $\alpha\beta\gamma = 0$. If $\gamma = \beta = 0$, the equation is $\alpha x^2 + 2\delta z + \epsilon = 0$, and the line L lies on the plane of symmetry $x = 0$. But this is the case of a parabolic cylinder and there is only one such plane. Hence, all lines of symmetry are lines of the plane $x = 0$. If the line L is defined by ·equations $x = 0$, $\lambda y + \mu z = p$, then the rotation of axes defined by

$$x' = x, \qquad y' = \lambda y + \mu z, \qquad z' = -\mu y + \lambda z$$

has as solved form

$$x = x', \qquad y = \lambda y' - \mu z', \qquad z = \mu y' + \lambda z'$$

and replaces the given equation by

$$\alpha x'^2 + 2\delta(\mu y' + \lambda z') + \epsilon = 0.$$

This rotation replaces the equation of L by $x' = 0$, $y' = p$. A translation of axes $x'' = x'$, $y'' = y' + p$, $z'' = z'$ yields

$$\alpha x''^2 + 2\delta[\mu(y'' + p) + \lambda z''] + \epsilon = 0.$$

This must be unaltered by the replacement of y'' by $-y''$. Then $\mu = 0$, and the line is given by $y = k = \pm p$. However, every line $x = 0$, $y = k$ is a line of symmetry of $\alpha x^2 + 2\delta z + \epsilon = 0$, since $x = y = 0$ is such a line, and the translation of axes $x' = x$, $y' = y + p$ does not alter the given equation. Interchanging the roles of y and z, we have the following results:

Theorem 11. *The lines of symmetry of the parabolic cylinder* $x^2 = \delta y$ *are all the lines of intersection of the plane* $x = 0$ *with planes* $z = k$ *perpendicular to the lines on the cylinder.*

If $\gamma = \alpha = 0$ and $\beta \neq 0$, we also obtain a parabolic cylinder. Thus the only remaining cases are $\gamma = \delta = 0$, $f(x, y, z) = \alpha x^2 + \beta y^2 + \epsilon$, and $\gamma \neq 0$, $f(x, y, z) = \alpha x^2 + \gamma z^2 + 2\delta z + \epsilon$, or $f(x, y, z) = \beta y^2 + \gamma z^2 + 2\delta z + \epsilon$. In any case we may reduce the equation to $\alpha x^2 + \beta y^2 + \epsilon = 0$ by a translation and rotation of axes and the line of symmetry becomes the line of intersection of two planes of symmetry. But, by Theorem 7, all planes of symmetry of a nonparabolic quadric cylinder intersect in a single line and we have proved the following:

Theorem 12. *A nonparabolic cylinder has a single line of symmetry.*

CHAPTER 8
SPHERICAL COORDINATES

1. Azimuth and elevation. It is sometimes convenient to locate points in space by three-dimensional vectors of coordinates called *spherical* coordinates. These coordinates have a position in space geometry like that of polar coordinates in plane geometry. They have the advantage over rectangular coordinates of being optically measureable from a fixed origin O of coordinates.

The first spherical coordinate is the *range r*. This is the distance

$$(1) \qquad r = |OP| = \sqrt{x^2 + y^2 + z^2}$$

from the origin O to the arbitrary point P in space, and we have given a formula expressing r in terms of the rectangular coordinates of P. We note that $r \geqq 0$ and that $r = 0$ only if P is at O. There are several physicogeometrical methods for the approximate measurement of r and, in particular, the measurement of the distance of a physical object from an origin of measurement is possible by radar devices.

The remaining two coordinates are angles α, ϵ measured (in radians) as in Fig. 16. The set of numbers (r, α, ϵ) is then a coordinate vector of three real numbers. If the angles are measured in degrees, they must be converted into radians. Actual measurement of α and ϵ may be achieved by a very simple and accurate instrument called a *surveyor's transit*.

The spherical coordinate angles α, ϵ may be defined most easily relative to a given rectangular coordinate system. We assume that $P = (x, y, z)$ is an arbitrary point in space and construct a half plane which contains P and has the z axis as its edge. In Fig. 16, this half plane intersects the x, y plane in a ray OA. Look down from the positive z direction on the x, y plane and so define a clockwise direction for angular measurement in the x, y plane. The angle α is then measured in a clockwise direction from the positive y axis to the ray OA. Evidently α

may be restricted to lie in the interval

(2) $$0 \leqq \alpha < 2\pi$$

radians.   The angle α is called the *azimuth* of P.

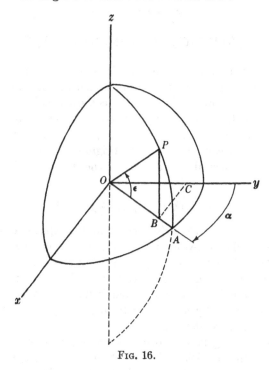

Fig. 16.

The angle ϵ is measured from the ray OA to the ray OP, and we may assume that

(3) $$-\frac{\pi}{2} \leqq \epsilon \leqq \frac{\pi}{2}.$$

Here $\epsilon > 0$ for points with $z > 0$ and $\epsilon \leqq 0$ otherwise.   We call ϵ the *elevation* of P.   Note that the x, y plane is the plane whose *spherical coordinate equation* is $\epsilon = 0$.

It is sometimes convenient to use an angle ζ called the *zenith* of P rather than the elevation angle ϵ.   This angle is measured from the positive z axis to the ray OP and $\zeta = \pi/2 - \epsilon$.   Then

$$0 \leqq \xi \leqq \pi.$$

Rectangular and spherical coordinates are related by the formulas

(4) $x = r \sin \alpha \cos \epsilon$, $y = r \cos \alpha \cos \epsilon$, $z = r \sin \epsilon$.

The last of these formulas comes from the fact that in Fig. 16 the distance $\overrightarrow{BP} = z$ and $|OP| = r$, so that $z = r \sin \epsilon$. Then $|OB| = r \cos \epsilon$ and $y = \overrightarrow{OC} = |OB| \cos \alpha$, and $x = \overrightarrow{CB} = |OB| \sin \alpha$. This verifies formula (4).

ORAL EXERCISES

1. The following vectors are the rectangular coordinates of points. Give their spherical coordinates.

(a) $(0, 1, 0)$ (d) $(0, 0, 5)$ (g) $(\sqrt{2}, \sqrt{6}, 2\sqrt{2})$
(b) $(1, 0, 0)$ (e) $(1, 1, 1)$ (h) $(\sqrt{6}, \sqrt{6}, -2)$
(c) $(0, 0, 1)$ (f) $(1, 1, 0)$ (i) $(-\sqrt{3}, 1, 2\sqrt{3})$

2. The following vectors are the spherical coordinates (r, α, ϵ) of corresponding points. Give their rectangular coordinates.

(a) $(0, \alpha, \epsilon)$ (d) $\left(2, \dfrac{\pi}{3}, \dfrac{\pi}{6}\right)$ (g) $\left(1, \dfrac{3\pi}{2}, -\dfrac{\pi}{4}\right)$

(b) $(1, 0, \epsilon)$ (e) $\left(3, \dfrac{\pi}{3}, \dfrac{\pi}{4}\right)$ (h) $\left(1, \dfrac{7}{6}\pi, -\dfrac{\pi}{4}\right)$

(c) $\left(1, \dfrac{\pi}{2}, \dfrac{\pi}{4}\right)$ (f) $\left(3, \pi, \dfrac{\pi}{4}\right)$ (i) $\left(4, \dfrac{7}{4}\pi, \dfrac{\pi}{4}\right)$

3. What geometrical object is described by the equation $\alpha = 0$? By the equation $\alpha(\alpha - \pi) = 0$?

2. The angle between two vectors. If P and P_0 are two points in space distinct from the origin O, we have already given a formula for the cosine of the angle $\theta = \angle POP_0$ in terms of the direction cosines of the rays OP and OP_0. We shall now derive a formula in terms of the spherical coordinates of P and P_0.

If $P = (r, \alpha, \epsilon)$ and $P_0 = (r_0, \alpha_0, \epsilon_0)$, the unit vector on the ray OP is $U = (1, \alpha, \epsilon)$ and that on the ray OP_0 is $U_0 = (1, \alpha_0, \epsilon_0)$. The rectangular coordinates of U are $(\sin \alpha \cos \epsilon, \cos \alpha \cos \epsilon, \sin \epsilon)$ and those of U_0 are $(\sin \alpha_0 \cos \epsilon_0, \cos \alpha_0 \cos \epsilon_0, \sin \epsilon_0)$, so that $\cos \theta = (\sin \alpha \sin \alpha_0 + \cos \alpha \cos \alpha_0) \cos \epsilon \cos \epsilon_0 + \sin \epsilon \sin \epsilon_0$. Then the required formula is

(5) $\cos \theta = \cos \epsilon \cos \epsilon_0 \cos (\alpha - \alpha_0) + \sin \epsilon \sin \epsilon_0.$

If θ is an angle so small that its sine is approximately equal to θ (measured in radians), we may replace formula (5) by a much simpler approximation formula. We shall use the notation

$$a \doteq b$$

(read a is approximately equal to b) throughout this chapter and the degree of accuracy of the approximation will of course depend on the actual approximations made and needs to be investigated in all specific cases.

Formula (5) implies that

$$1 - \cos\theta = \cos\epsilon\cos\epsilon_0[1 - \cos(\alpha - \alpha_0)] + 1$$
$$(6) \qquad\qquad\qquad\qquad - \cos\epsilon\cos\epsilon_0 - \sin\epsilon\sin\epsilon_0$$
$$= \cos\epsilon\cos\epsilon_0[1 - \cos(\alpha - \alpha_0)] + 1 - \cos(\epsilon - \epsilon_0).$$

If ϕ is any angle, the formula $1 - \cos\phi = 2\sin^2\phi/2$ is a well-known formula of elementary trigonometry. We apply it to formula (6) to obtain

$$\sin^2\frac{\theta}{2} = \sin^2\left(\frac{\alpha - \alpha_0}{2}\right)\cos\epsilon\cos\epsilon_0 + \sin^2\left(\frac{\epsilon - \epsilon_0}{2}\right),$$

and therefore have the exact formula

$$(7) \qquad \sin\frac{\theta}{2} = \sqrt{\sin^2\left(\frac{\alpha - \alpha_0}{2}\right)\cos\epsilon\cos\epsilon_0 + \sin^2\left(\frac{\epsilon - \epsilon_0}{2}\right)}$$

for acute angles θ. When θ is small, the angles $\alpha - \alpha_0$ and $\epsilon - \epsilon_0$ are small and $\cos\epsilon \doteq \cos\epsilon_0$, $\sin\frac{1}{2}(\alpha - \alpha_0) \doteq \frac{1}{2}(\alpha - \alpha_0)$, $\sin\frac{1}{2}(\epsilon - \epsilon_0) \doteq \frac{1}{2}(\epsilon - \epsilon_0)$. Then formula (7) becomes

$$(8) \qquad\qquad \theta \doteq \sqrt{[(\alpha - \alpha_0)\cos\epsilon]^2 + (\epsilon - \epsilon_0)^2}.$$

EXERCISES

1. Compute θ in milliradians (thousands of radians) in the following cases by the use of the exact formula (7) and five-place tables of logarithms. Give your answer correct to the nearest tenth of a milliradian.

(a) $\alpha = 17°27'$, $\alpha_0 = 18°42'$, $\epsilon = 15°$, $\epsilon_0 = 16°$

(b) $\alpha = 24°32'$, $\alpha_0 = 24°$, $\epsilon = 72°$, $\epsilon_0 = 69°40'$

(c) $\alpha = 192°17'$, $\alpha_0 = 191°$, $\epsilon = -18°$, $\epsilon_0 = -19°$

2. Compute θ in milliradians correct to the nearest tenth of a milliradian by the use of formula (8), and compare the results with those of Exercise 1.

3. Parallax. The changes in spherical coordinates that are the results of a *translation* of axes are called *parallax* corrections. There are many physical situations where measurements of range, azimuth, and elevation are made by instruments located at a point O, but are required relative to a parallel coordinate system with origin at a nearby point O'. In such cases, formulas for parallax corrections are needed.

Let us suppose that the x, y, z coordinates of O' are (a, b, c) and that the corresponding spherical coordinates are d, α_0, ϵ_0. Thus

$$(9) \quad d = \sqrt{a^2 + b^2 + c^2}, \quad \sin \epsilon_0 = \frac{c}{d}, \quad \tan \alpha_0 = \frac{a}{b}.$$

Suppose then that $P = (r, \alpha, \epsilon)$ has rectangular coordinates x, y, z, and define

$$(10) \qquad \delta = \frac{d}{r}, \quad \cos \theta = \frac{ax + by + cz}{dr},$$

so that θ is the angle between the ray OO' and the ray OP. Then

$$(11) \qquad d \cos \theta = (a \sin \alpha + b \cos \alpha) \cos \epsilon + c \sin \epsilon,$$

and $d \cos \theta$ may be computed when α and ϵ are given. The numbers a, b, c may be regarded as fixed, and it is possible to construct a table of corresponding values of $d \cos \theta$ in which the table is entered with α and ϵ and $d \cos \theta$ is read from the table. We may, of course, apply formula (7) to compute θ and then compute $d \cos \theta$.

When many parallax corrections to the same point O' must be made, it is worth while to construct a table as described above and to translate it into a *contour graph*. This graph will consist of a number of curves. Each curve will consist of points whose x coordinates are α radians and y coordinates are ϵ radians, and $d \cos \theta$ will be the same for *all points on the same* curve. If enough curves are drawn, it is possible to read off $d \cos \theta$ directly from the graph by finding the curve on which a point with prescribed coordinates lies. Linear interpolation between curves is sufficiently accurate, of course, when many curves are drawn. It is usually possible to draw the curves by plotting a relatively small number of points on each curve.

Let us suppose now that $d \cos \theta$ can be quickly determined by a table or a contour graph whenever α and ϵ are given. We require the range r', the azimuth α', and the elevation ϵ' of the point P relative to the coordinate system obtained by a translation of axes carrying the origin to O'. Then the parallax corrections are defined to be

$$(12) \quad \Delta(r) = r' - r, \qquad \Delta(\alpha) = \alpha' - \alpha, \qquad \Delta(\epsilon) = \epsilon' - \epsilon,$$

and we propose to find formulas for the computation of $\Delta(r)$, $\Delta(\alpha)$, $\Delta(\epsilon)$.

We use the translation formulas $x' = x - a$, $y' = y - b$, $z' = z - c$, and compute $r'^2 = x'^2 + y'^2 + z'^2 = x^2 + y^2 + z^2 - 2(ax + by + cz) + a^2 + b^2 + c^2 = r^2 - 2dr \cos \theta + d^2$ by formula (10). Then

$$(13) \quad r'^2 = r^2(1 - 2\delta \cos \theta + \delta^2) = r^2[(1 - \delta \cos \theta)^2 + (\delta \sin \theta)^2].$$

The assumption that d is small compared with r implies that $1 > \delta \cos \theta$,

$$(14) \qquad r' = r(1 - \delta \cos \theta) \sqrt{1 + \left(\frac{\delta \sin \theta}{1 - \delta \cos \theta}\right)^2}.$$

Expand this expression by the binomial theorem which states that

$$(1 + b)^n = 1 + nb + \frac{n(n - 1)}{2} b^2 + \cdots$$

and which yields a convergent series if $|b| < 1$ and $n = \frac{1}{2}$. Then

$$(15) \quad r' = r(1 - \delta \cos \theta) \left[1 + \frac{\delta^2 \sin^2 \theta}{2(1 - \delta \cos \theta)^2} - \frac{\delta^2 \sin^4 \theta}{8(1 - \delta \cos \theta)^4} + \cdots \right]$$

and therefore $\delta r = d$ yields

$$(16) \qquad r' \doteq r - d \cos \theta + \frac{d^2 \sin^2 \theta}{2(r - d \cos \theta)}.$$

The parallax correction

$$(17) \qquad \qquad \Delta(r) \doteq -d \cos \theta$$

is usually quite adequate and may be computed as was indicated above. The more accurate correction

$$(18) \qquad \qquad \Delta(r) \doteq -d \cos \theta + \frac{d^2 - d^2 \cos^2 \theta}{2(r - d \cos \theta)}$$

may also be computed easily from given values of $d \cos \theta$, but the difficulty in making exceedingly accurate measurements of r usually make this more refined correction meaningless.

We next observe that $r' \sin \epsilon' = z' = z - c = r \sin \epsilon - c$. This gives

$$
(19) \qquad \sin \epsilon' = \frac{r \sin \epsilon - c}{r'},
$$

and

$$
(20) \qquad \sin \epsilon' - \sin \epsilon = \frac{(r - r') \sin \epsilon - c}{r'}.
$$

Then we use a well-known formula of trigonometry to replace formula (20) by

$$
(21) \qquad 2 \sin \frac{\Delta \epsilon}{2} \cos \frac{\epsilon + \epsilon'}{2} = \frac{(r - r') \sin \epsilon - c}{r'}.
$$

If a, b, c are small compared with r; the values of $\Delta(\epsilon)$ will be small, and we may use the approximations

$$
(22) \qquad
\begin{aligned}
\sin \frac{\Delta(\epsilon)}{2} &\doteq \frac{\Delta(\epsilon)}{2}, \\
\cos \frac{\epsilon + \epsilon'}{2} &= \cos \left(\epsilon + \frac{\Delta(\epsilon)}{2} \right) \doteq \cos \epsilon,
\end{aligned}
$$

and thus obtain

$$
(23) \qquad -\Delta(\epsilon) \doteq \frac{\Delta(r)}{r'} \tan \epsilon + \frac{c}{r' \cos \epsilon}.
$$

This correction is proportional to r' and there is little loss of accuracy if the approximation

$$
(24) \qquad \Delta(\epsilon) \doteq \frac{1}{r} \frac{d \cos \theta \sin \epsilon - c}{\cos \epsilon}
$$

is used. The answer is the number of radians in $\Delta(\epsilon)$ and is converted to milliradians by multiplication by 1,000.

As in the case of the computation of $\Delta(r)$, the computation of many values of $\Delta(r)$ for fixed a, b, c may be achieved best by a graph of curves giving equal values of $\Delta(\epsilon)$ for $r = 1$ and α and ϵ as variables. The computation of $\Delta(r)$ is then a matter of reading a result from a graph and dividing the result by r.

To derive a formula for $\Delta(\alpha)$, we note that

$$
\tan \alpha' = \frac{x'}{y'} = \frac{r \sin \alpha \cos \epsilon - a}{r \cos \alpha \cos \epsilon - b}.
$$

Then

$$\tan \alpha' - \tan \alpha =$$
$$\frac{(r \sin \alpha \cos \epsilon - a) \cos \alpha - (r \cos \alpha \cos \epsilon - b) \sin \alpha}{\cos \alpha (r \cos \alpha \cos \epsilon - b)}.$$

This formula reduces to

$$\frac{\sin (\alpha' - \alpha)}{\cos \alpha \cos \alpha'} = \frac{a \cos \alpha + b \sin \alpha}{\cos \alpha (r \cos \alpha \cos \epsilon - b)}.$$

It follows that

$$(25) \qquad \sin (\alpha' - \alpha) = \cos \alpha' \frac{a \cos \alpha + b \sin \alpha}{r \cos \alpha \cos \epsilon - b}.$$

We make the approximations $\sin (\alpha' - \alpha) \doteq \alpha' - \alpha = \Delta(\alpha)$, $\cos \alpha' \doteq \cos \alpha$, to obtain

$$(26) \qquad \Delta(\alpha) \doteq \frac{a \cos \alpha + b \sin \alpha}{r \cos \epsilon - b \sec \alpha}.$$

But then

$$\Delta(\alpha) \doteq \frac{a \cos \alpha + b \sin \alpha}{r \cos \epsilon}$$
$$+ (a \cos \alpha + b \sin \alpha) \left(\frac{1}{r \cos \epsilon - b \sec \alpha} - \frac{1}{r \cos \epsilon} \right)$$
$$\doteq \frac{a \cos \alpha + b \sin \alpha}{r \cos \epsilon} 1 + \left(\frac{b \sec \alpha}{r \cos \epsilon - b \sec \alpha} \right)$$
$$= \frac{a \cos \alpha + b \sin \alpha}{r \cos \epsilon} \left(1 + \frac{b}{r \cos \epsilon \cos \alpha - b} \right).$$

If b/r is small, we may use the approximation

$$(27) \qquad \Delta(\alpha) \doteq \frac{a \cos \alpha + b \sin \alpha}{r \cos \epsilon}$$

and may compute $\Delta(\alpha)$ by a process similar to that used in the computation of $\Delta(\epsilon)$. Otherwise, it will be necessary to multiply the result by the factor

$$(28) \qquad \frac{\beta}{\cos \alpha \cos \epsilon - \beta}$$

where $\beta = b/r$.

EXERCISES

1. Let $a = 2$ yards, $b = -3$ yards, and $c = 6$ yards. Construct a table of values of $d \cos \theta$ to tenths of a yard for $\alpha = 70°$, $71°$, $72°$, $73°$, $74°$ and $\epsilon = 15°$, $16°$, $17°$, $18°$, $19°$.

2. Compute the values of $\Delta(r)$ by the use of formula (17) for the following values of (r, α, ϵ), where we are giving α and ϵ in degrees and r in yards.

(a) $(1,000, 70°, 18°)$ (d) $(300, 73°, 15°)$
(b) $(972, 71°, 17°)$ (e) $(100, 72°, 19°)$
(c) $(500, 74°, 15°)$ (f) $(75, 71°, 18°)$

3. Compute $\Delta(r)$ to tenths of a yard by the use of formula (18), and compare results.

4. Compute $\Delta(\epsilon)$ in milliradians by the use of formula (24) and compare the result with that obtained by replacing r by r'.

5. Compare $\Delta(\alpha)$ by the use of formula (27) for parts (a) and (b) of Exercise 2. Compute the correction factor of formula (28), and compare results.

4. Other spherical coordinates. The spherical coordinates α, ϵ described in Sec. 1 are defined for every point P in terms of a half plane through P and the z axis. It should seem natural then to define sets of corresponding angles relative to the remaining coordinate axes.

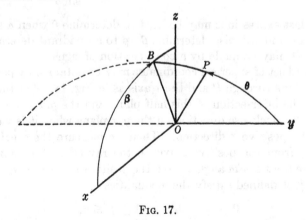

Fig. 17.

We first construct a half plane through P and the y axis as in Fig. 17. Let this half plane cut the x, z plane in a ray OB and define the clockwise direction on the x, z plane as viewed from the positive y direction. Then we measure the y *azimuth* angle β in this clockwise direction from the positive x axis to the ray OB and

(29) $$0 \leqq \beta < 2\pi$$

radians. It follows that $0 \leqq \beta \leqq \pi$ for points having positive or zero elevation and $\pi < \beta < 2\pi$ for points having negative elevation.

An angle θ may also be measured from the positive y axis to the ray OP. Then

(30) $$0 \leqq \theta \leqq \pi,$$

and the angle θ is a zenith angle rather than an elevation angle. We shall call θ the y *zenith* of P. It is called the *angle off the nose* in aerial navigation.

The set of spherical coordinates just defined is determined by the relations

(31) $x = r \sin \theta \cos \beta,$ $y = r \cos \theta,$ $z = r \sin \theta \sin \beta.$

We leave the verification of this formula to the reader. It may be combined with formula (4) to yield

(32) $$\cos \theta = \cos \alpha \cos \epsilon, \qquad \tan \beta = \frac{\tan \epsilon}{\sin \alpha}.$$

These last expressions may be used to determine θ when α and ϵ are given and will also determine β up to a quadrant determination that may be made by a consideration of signs.

A third set of spherical coordinates may be obtained by passing a half plane through P and the x axis as in Fig. 18. Let the ray OD be the intersection of this half plane and the y, z plane, and define the clockwise direction on the y, z plane when it is viewed from the positive x direction. Then we measure the x *azimuth* angle γ from the positive z axis to the ray OD. We similarly measure the x *zenith* angle ψ from the positive x axis to OQ. The angles just defined satisfy the inequalities

(33) $$0 \leqq \gamma < 2\pi, \qquad 0 \leqq \psi \leqq \pi.$$

They also satisfy the relations

(34) $x = r \cos \psi,$ $y = r \sin \psi \cos \gamma,$ $z = r \sin \psi \sin \gamma,$

and γ and ψ may be determined when α and ϵ are given by the formulas

(35) $\cos \psi = \sin \alpha \cos \epsilon,$ $\tan \gamma = \dfrac{\tan \epsilon}{\cos \alpha}$

and a consideration of signs.

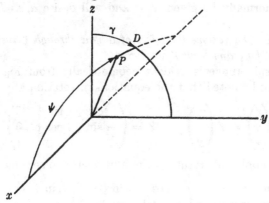

Fig. 18.

5. The matrices of planar rotations. We have called a rotation of axes in which one coordinate axis remains unaltered a *planar* rotation. If the unaltered axis is the z axis, we call the

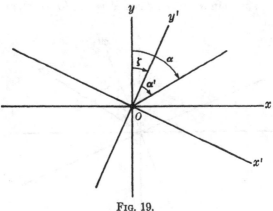

Fig. 19.

rotation a *yaw*. We call the rotation a *pitch* if the unaltered axis is the x axis and a *roll* if the unaltered axis is the y axis. We shall measure all three of these angles in a clockwise positive direction as viewed from the positive direction on the unaltered axis on the coordinate plane perpendicular to this axis.

In Fig. 19 a yaw through an angle ζ has been performed that carries x, y, z to x', y', z'. It should be evident that ϵ is unaltered and that

$$\alpha = \alpha' + \zeta, \qquad \epsilon = \epsilon'.$$

We will normally be given α', ϵ' and will desire α, ϵ. Then we have

RULE I. *To remove the effect of a yaw through ζ increase the azimuth α' by ζ and leave ϵ' unaltered.*

The result above is evident geometrically from Fig. 19. It should also be noted that the equations of rotation are

$$(36) \qquad \begin{pmatrix} x \\ y \\ z \end{pmatrix} = Y \begin{pmatrix} x' \\ y' \\ z' \end{pmatrix}, \qquad Y = \begin{pmatrix} \cos \zeta & \sin \zeta & 0 \\ -\sin \zeta & \cos \zeta & 0 \\ 0 & 0 & 1 \end{pmatrix},$$

where we apply this result to the unit vector to see that

$$(37) \qquad \begin{pmatrix} \sin \alpha \cos \epsilon \\ \cos \alpha \cos \epsilon \\ \sin \epsilon \end{pmatrix} = \begin{pmatrix} \cos \zeta & \sin \zeta & 0 \\ -\sin \zeta & \cos \zeta & 0 \\ 0 & 0 & 1 \end{pmatrix} \begin{pmatrix} \sin \alpha' \cos \epsilon' \\ \cos \alpha' \cos \epsilon' \\ \sin \epsilon' \end{pmatrix}.$$

This formula implies that $\sin \epsilon = \sin \epsilon'$ and so $\epsilon = \epsilon'$, $\cos \epsilon = \cos \epsilon'$. Hence, $\sin \alpha = \sin \alpha' \cos \zeta + \cos \alpha' \sin \zeta = \sin (\alpha' + \zeta)$, $\cos \alpha = -\sin \zeta \sin \alpha' + \cos \zeta \cos \alpha' = \cos (\alpha' + \zeta)$, and so $\alpha = \alpha' + \zeta$.

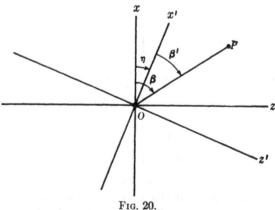

FIG. 20.

In a similar fashion we use Fig. 20 to measure a roll angle η about the y axis. Then $\beta = \beta' + \eta$, $\theta' = \theta$, and we have the following rule:

Rule II. *To remove the effect of a roll through our angle η increase the y azimuth β' by η, and leave the y zenith θ' unaltered.*

As before, the equations of rotation are

$$(38) \quad \begin{pmatrix} x \\ y \\ z \end{pmatrix} = R_\eta \begin{pmatrix} x' \\ y' \\ z' \end{pmatrix}, \qquad R_\eta = \begin{pmatrix} \cos \eta & 0 & -\sin \eta \\ 0 & 1 & 0 \\ \sin \eta & 0 & \cos \eta \end{pmatrix}$$

where we use

$$(39) \quad \begin{pmatrix} \sin \theta \cos \beta \\ \cos \theta \\ \sin \theta \sin \beta \end{pmatrix} = \begin{pmatrix} \cos \eta & 0 & -\sin \eta \\ 0 & 1 & 0 \\ \sin \eta & 0 & \cos \eta \end{pmatrix} \begin{pmatrix} \sin \theta' \cos \beta' \\ \cos \theta' \\ \sin \theta' \sin \beta' \end{pmatrix}$$

to verify that $\theta = \theta'$, $\cos \beta = \cos \eta \cos \beta' - \sin \eta \sin \beta' = \cos (\eta + \beta')$, $\sin \beta' = \sin \eta \cos \beta' + \cos \eta \sin \beta' = \sin (\eta + \beta')$, $\beta = \eta + \beta'$.

We finally use Fig. 21 to measure a pitch angle ξ such that

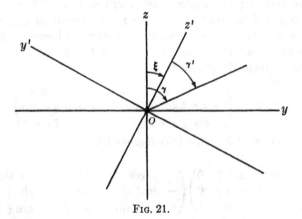

Fig. 21.

$\gamma = \gamma' + \xi, \psi = \psi'$. Then we have the following rule:

Rule III. *To remove the effect of a pitch through an angle ξ increase the x azimuth γ' by ξ and leave the x zenith ψ unaltered.*

The equations of rotation in this case are

$$(40) \quad \begin{pmatrix} x \\ y \\ z \end{pmatrix} = P_\xi \begin{pmatrix} x' \\ y' \\ z' \end{pmatrix}, \qquad P_\xi = \begin{pmatrix} 1 & 0 & 0 \\ 0 & \cos \xi & -\sin \xi \\ 0 & \sin \xi & \cos \xi \end{pmatrix},$$

where we verify the correctness of our formulas by the computation

$$(41) \quad \begin{pmatrix} \cos \psi \\ \sin \psi \cos \gamma \\ \sin \psi \sin \gamma \end{pmatrix} = \begin{pmatrix} 1 & 0 & 0 \\ 0 & \cos \xi & -\sin \xi \\ 0 & \sin \xi & \cos \xi \end{pmatrix} \begin{pmatrix} \cos \psi \\ \sin \psi' \cos \gamma' \\ \sin \psi' \sin \gamma' \end{pmatrix}.$$

This formula yields $\psi = \psi'$, $\cos \gamma = \cos \xi \cos \gamma' - \sin \xi \sin \gamma'$ $= \cos (\xi + \gamma')$, $\sin \gamma = \sin \xi \cos \gamma' + \cos \xi \sin \gamma' = \sin (\xi + \gamma')$.

6. Rotations as products of planar rotations. In Sec. 3 of Chap. 7 we gave a geometrical argument showing that every rotation of axes may be expressed as a product of three planar rotations. There are *many* expressions of a rotation of axes in terms of three planar rotations, and in particular we may write

$$(42) \quad \begin{pmatrix} x \\ y \\ z \end{pmatrix} = L \begin{pmatrix} x' \\ y' \\ z' \end{pmatrix}, \qquad L = \begin{pmatrix} \lambda_1 & \lambda_2 & \lambda_3 \\ \mu_1 & \mu_2 & \mu_3 \\ \nu_1 & \nu_2 & \nu_3 \end{pmatrix} = Y_\zeta P_\xi R_\eta.$$

Then we have expressed the rotation matrix L as a product of a yaw matrix, a pitch matrix, and a roll matrix. There actually exist physical instruments for the measurement of ζ, ξ, η. We shall be interested here primarily in observing how the elements of the matrix L are related to the angles ξ, η, ζ.

We first form the product

$$(43) \quad P_\xi R_\eta = \begin{pmatrix} \cos \eta & 0 & -\sin \eta \\ -\sin \xi \sin \eta & \cos \xi & -\sin \xi \cos \eta \\ \cos \xi \sin \eta & \sin \xi & \cos \xi \cos \eta \end{pmatrix}.$$

The definition of formula (42) implies that

$$(44) \quad L =$$
$$\begin{pmatrix} \cos \zeta & \sin \zeta & 0 \\ -\sin \zeta & \cos \zeta & 0 \\ 0 & 0 & 1 \end{pmatrix} \begin{pmatrix} \cos \eta & 0 & -\sin \eta \\ -\sin \xi \sin \eta & \cos \xi & -\sin \xi \cos \eta \\ \cos \xi \sin \eta & \sin \xi & \cos \xi \cos \eta \end{pmatrix},$$

and it follows that

$$(45) \quad \begin{aligned}
\lambda_1 &= \cos \zeta \cos \eta - \sin \zeta \sin \xi \sin \eta, \\
\lambda_2 &= \sin \zeta \cos \xi, \\
-\lambda_3 &= \cos \zeta \sin \eta + \sin \zeta \sin \xi \cos \eta, \\
\mu_1 &= -\sin \zeta \cos \eta - \cos \zeta \sin \xi \sin \eta, \\
\mu_2 &= \cos \zeta \cos \xi, \\
\mu_3 &= \sin \eta \sin \xi - \cos \zeta \sin \xi \cos \eta, \\
\nu_1 &= \cos \xi \sin \eta, \qquad \nu_2 = \sin \xi, \qquad \nu_3 = \cos \xi \cos \eta.
\end{aligned}$$

When L is given, the angles ξ, η, ζ may be determined, apart from quadrant, by the use of the formulas

$$(46) \qquad \sin \xi = \nu_2, \qquad \tan \eta = \frac{\nu_1}{\nu_3}, \qquad \tan \zeta = \frac{\lambda_2}{\mu_2}.$$

In the case where all rotation angles lie between $-\pi/2$ and $\pi/2$, formula (46) completely determines ξ, η, ζ.

It should be clear that if we write L as a product of three planar rotations in a different order, then the yaw pitch and roll angles are different; for example, if $L = Y_\zeta P_\xi R_\eta = P_\lambda R_\mu Y_\nu$, then, in general, it will not be true that $\zeta = \nu$.

7. Stabilization of coordinates. The problem of stabilizing coordinates arises in physical situations where it is required to find the azimuths α and elevations ϵ of many points P relative to a fixed coordinate system under conditions where the azimuth α' and elevation ϵ' of each point P relative to a rotated coordinate system, as well as the corresponding rotation, are measurable. It is customary to call α', ϵ' the *unstabilized coordinates* of P and α, ϵ the *stabilized coordinates* of P and we are thinking of a case where the rotation varies as P changes.

Let us assume that the rotation of axes is given by an equation

$$(47) \qquad \begin{pmatrix} \sin \alpha \cos \epsilon \\ \cos \alpha \cos \epsilon \\ \sin \epsilon \end{pmatrix} = Y_\zeta P_\xi R_\eta \begin{pmatrix} \sin \alpha' \cos \epsilon' \\ \cos \alpha' \cos \epsilon' \\ \sin \epsilon' \end{pmatrix},$$

where we have written the effect on the general unit vector of a rotation which is the product of a yaw, a pitch, and a roll. Since

$$(48) \qquad Y_\zeta^{-1} = Y_{-\zeta} = \begin{pmatrix} \cos \zeta & -\sin \zeta & 0 \\ \sin \zeta & \cos \zeta & 0 \\ 0 & 0 & 1 \end{pmatrix},$$

we see that

$$(49) \qquad \begin{pmatrix} \sin (\alpha - \zeta) \cos \epsilon \\ \cos (\alpha - \zeta) \cos \epsilon \\ \sin \epsilon \end{pmatrix} = P_\xi R_\eta \begin{pmatrix} \sin \alpha' \cos \epsilon' \\ \cos \alpha' \cos \epsilon' \\ \sin \epsilon' \end{pmatrix}.$$

The required coordinates may then be obtained by the following procedure:

  a. Convert α', ϵ' to β', θ' by the use of formula (32).
  b. Compute $\beta'' = \beta + \eta$, $\theta'' = \theta'$.

    c. Convert β'', θ'' to γ'', ψ'' by the use of the formulas

$$\cos \psi = \sin \theta \cos \beta, \qquad \tan \gamma = \tan \theta \sin \beta.$$

    d. Compute $\gamma''' = \gamma'' + \xi$, $\psi''' = \psi''$.

    e. Convert γ''', ψ''' to α''', ϵ''' by the use of the formulas

$$\sin \epsilon = \sin \psi \sin \gamma, \qquad \tan \alpha = \cot \psi \sec \gamma.$$

    f. Then $\alpha = \zeta + \alpha'''$, $\epsilon = \epsilon'''$.

The formulas given in (*c*) and (*d*) are obtained by combining formulas (4), (31), and (35). The procedure above evidently requires a considerable amount of computation.

A more direct formula is obtained by the use of formula (42) and matrix multiplication. It yields

(50) $\sin (\alpha - \zeta) \cos \epsilon = \cos \eta \sin \alpha' \cos \epsilon' - \sin \eta \sin \epsilon'$

(51) $\cos (\alpha - \zeta) \cos \epsilon = -\sin \xi \sin \eta \sin \alpha' \cos \epsilon'$
$$+ \cos \xi \cos \alpha' \cos \epsilon' - \sin \xi \cos \eta \sin \epsilon'$$

(52) $\sin \epsilon = \cos \xi \sin \eta \sin \alpha' \cos \epsilon' + \sin \xi \cos \alpha' \cos \epsilon'$
$$+ \cos \xi \cos \eta \sin \epsilon'.$$

The computation of α and ϵ by the use of these formulas should require less computation than the procedure outlined above. However, there is a graphical (gnomonic chart) method for using the procedure above that makes it a very rapid one.

If the rotation angles ξ, η, ζ are very small, we can use the approximations

$$\sin \eta \doteq \eta, \qquad \sin \xi \doteq \xi, \qquad \cos \eta \doteq \cos \xi = 1,$$
$$\sin \xi \sin \eta \doteq 0.$$

Then formulas (50), (51), and (52) yield

(53) $\tan (\alpha - \zeta) \doteq \dfrac{(\sin \alpha' - \eta \cos \alpha') \cos \epsilon'}{\cos \alpha' \cos \epsilon' - \zeta \sin \epsilon'}$

(54) $\sin \epsilon \doteq (\eta \sin \alpha' + \xi \cos \alpha') \cos \epsilon' + \sin \epsilon'.$

We may then write $\sin \epsilon - \sin \epsilon' = 2 \sin \dfrac{\epsilon - \epsilon'}{2} \cos \dfrac{\epsilon + \epsilon'}{2} \doteq$

$(\epsilon - \epsilon') \cos \epsilon'$ and thus obtain the further approximation

(55) $\epsilon = \epsilon' + \eta \sin \alpha' + \xi \cos \alpha'$

for a rapid computation of ϵ. However, even these approximations are not so rapid as the graphical method we have referred to above.

8. Gnomonic projections. The *gnomonic projection* of a point P on the plane G whose rectangular coordinate equation is $y = a > 0$ is the point Q of G whose azimuth and elevation are those of P. The plane G is tangent to the sphere of radius a at the point of zero azimuth and elevation, and Q is the point of intersection of the line $x = t \sin \alpha \cos \epsilon, y = t \cos \alpha \cos \epsilon, z = t \sin \epsilon$ with G. Then the coordinates of Q are

$$(56) \qquad x = a \tan \alpha, \qquad y = a, \qquad z = a \tan \epsilon \sec \alpha.$$

The gnomonic projection *of any surface* S on G is the locus of all the points of G that are the gnomonic projections of the points of S. Consider in particular the cone consisting of all points in space having a fixed elevation ϵ. The projections of these points are given by formula (56) for ϵ fixed, and they satisfy $z^2 = a^2 \tan^2 \epsilon \sec^2 \alpha = \tan^2 \epsilon(a^2 \tan^2 \alpha + a^2) = \tan^2 \epsilon(x^2 + a^2)$. Then the gnomonic projection of the cone of points having elevation $\epsilon = 0$ is the hyperbola

$$(57) \qquad \frac{z^2}{\tan^2 \epsilon} - x^2 = a^2, \qquad y = a.$$

The points of zero elevation project into the line $z = 0, y = a$.

The points having a fixed azimuth α lie on a half plane that is a part of the plane $x = y \tan \alpha$. Then the locus of the gnomonic projections of these points is the vertical line

$$(58) \qquad x = a \tan \alpha, \qquad y = a.$$

To obtain the gnomonic projections of those points which have a constant y zenith angle θ, we use formula (31) and put $y = a$. Then

$$(59) \quad x = a \tan \theta \cos \beta, \qquad y = a, \qquad z = a \tan \theta \sin \beta,$$

so that

$$(60) \qquad x^2 + z^2 = a^2 \tan^2 \theta, \qquad y = a.$$

Hence, the projection of the cone of all points having y zenith equal to θ lie on a circle of G with center at $(0, a, 0)$ and radius $a \tan \theta$. The points having y azimuth β all lie on the plane $z = x \tan \beta$, and this plane cuts the plane G in the line

$$(61) \qquad z = x \tan \beta, \qquad y = a.$$

Here β is the angle from the coordinate axis $z = 0$, $y = a$ in G to this line as measured in the usual counterclockwise direction. Note that G is parallel to the zx plane and that we are viewing G from the negative y direction.

Let us finally consider the x zenith and azimuth. We use formula (34) to write

$$(62) \qquad x = a \cot \psi \sec \gamma, \qquad y = a, \qquad z = a \tan \gamma.$$

Then the gnomonic projections of all points whose x zenith is ψ satisfy the equation $x^2 = a^2 \cot^2 \psi \sec^2 \gamma = \cot^2 \psi (a^2 \tan^2 \gamma + a^2) = \cot^2 \psi (z^2 + a^2)$. Thus if $\psi \neq \pi/2$, the locus of the gnomonic projections of the points having x zenith ψ is the hyperbola

$$(63) \qquad \frac{x^2}{\cot^2 \psi} - z^2 = a^2, \qquad y = a.$$

The points with $\psi = \pi/2$ are the points whose projections from the line $x = 0$, $y = a$, and these are the points with $\alpha = 0$.

Finally, the points having fixed x azimuth γ lie on the horizontal line

$$z = a \tan \gamma, \qquad y = a.$$

We have shown how we may determine curves on G that, when properly labeled, yield the six spherical coordinates of all points in space.

Gnomonic projections may be made on other planes than the one we have selected, but are always made on planes perpendicular to a radius of a sphere with center at the origin. The projections on planes parallel to the z axis are all called *equatorial* projections, and those on planes perpendicular to the z axis are called *polar* projections.

EXERCISES

1. Derive the curves of gnomonic projection for points having a constant spherical coordinate α, ϵ, β, 0, γ, or ψ on the following planes G, where $a > 0$:

(a) $y = -a$ (c) $x = -a$
(b) $x = a$ (d) $z = a$

2. Derive the equations of the curve of gnomonic projection of all points of elevation ϵ on the plane perpendicular to the line $x = y$, $z = 0$.

9. Gnomonic charts.   A gnomonic chart is a map of a portion of a plane that is tangent to a sphere of radius a inches. The value of the constant a depends on the reading accuracy desired.

We take the origin of a rectangular coordinate system on the chart to be the point $C = (0, a, 0)$ and the chart to be the plane $y = a$. Then C may also be taken to be the origin of a translated (x', y', z') coordinate system in which the tangent plane at C is the x', z' plane, *i.e.*, the plane $y' = 0$. The x' axis is then the gnomonic projection of all points having elevation zero and is represented on the chart by a horizontal line. The z' axis is the gnomonic projection of all points having azimuth zero and is represented on the chart by a vertical line. All points having azimuth $\alpha = A°$ project into a vertical line and thus the azimuth of the projection P' on the chart of a point P in space is the same as the azimuth of P and may be read by the use of a scale on the x' axis in which the point marked $A°$ is at the distance $x' = a \tan A$ from C. The reading of the coordinate A may be facilitated by the printing on the chart of a vertical grid. It is necessary for the finite portion of the tangent plane represented by the chart to include the portion where $-k° \leqq \alpha \leqq k°$, where k is somewhat more than 45. Then the chart extends to a point where x' is greater than a inches. Since very large charts are difficult to handle, there are definite physical limitations on the use of gnomonic charts as accurate devices for the reading of spherical coordinates.

All points having a fixed value of the coordinate γ project into a horizontal line and therefore may be read by the use of a scale on the z' axis in which the point marked $E°$ is at a distance of $z' = a \tan E°$ from C. The spherical coordinate β could be read by the use of a circular protractor scale on the chart and a rotating arm pivoted at C to read the coordinate $\beta = B°$. The arm could contain a scale like the A scale or it could have a slider that would be used to locate the point P' accurately. The arm could then be rotated to either the A scale or the E scale for the reading of the coordinate θ.

All points having the same elevation ϵ lie on a cone intersecting the tangent plane in a hyperbola. A gnomonic chart then contains a quasihorizontal family of hyperbolas. The hyperbola that cuts the z' axis in the point labeled $E°$ has the rectangular coordinate equation $z = \tan E° \sqrt{x^2 + a^2}$, $y = a$, and all points whose gnomonic projections lie on this hyperbola have elevation $\epsilon = E°$. The number of such curves which are actually drawn depends partly on the reading accuracy desired and partly on the physical limitations which the drawing of accurate curves and readability will impose. The chart must also contain a family of quasivertical hyperbolas whose equations are $x = \tan A° \sqrt{z^2 + a^2}$, $y = a$, and all the points whose projections lie on the hyperbola cutting the x' axis in the point labeled $A°$ have the same x zenith $\psi = 90° - A°$.

We have now seen how a gnomonic chart may be used as a direct reading device for the six spherical coordinates α, β, ψ, ϵ, θ, γ of any point on the chart. Thus, if any two of these coordinates are given for any point P in space, a corresponding chart projection P' may be plotted and the remaining coordinates of P may be read. The chart may be used also as a device for computing the effect on α, ϵ of a rotation of axes when the rotation has been expressed as a product of planar rotations. For the rules given in Sec. 5 may be applied and the operations of that section become chart motions.

In using gnomonic charts, the limitation of a chart to a finite portion of the tangent plane requires that the charts be interpreted also as tangent planes whose point of tangency is one of the points $(a, 0, 0)$, $(-a, 0, 0)$, $(0, -a, 0)$, $(0, 0, a)$, $(0, 0, -a)$ as well as the point $(0, a, 0)$ used in the description above. The symmetry of our definitions of the three pairs of spherical coordinates implies that the six spherical coordinates may be obtained by making the same readings as before. However, the correspondence between readings and coordinates will be different for each case.

The interested student should construct a chart with $a = 10$ inches and curves drawn for every three degrees of A and E so as to get a better idea of the use of such a chart.

EXERCISES

1. Derive the interpretations of the six angles measured on a gnomonic chart for each of the positions of C.

2. Interpret planar rotations as chart motions in the case where $C = (0, a, 0)$. Give the interpretations also for $C = (0, -a, 0)$, $C = (a, 0, 0)$, and $C = (0, 0, a)$.

CHAPTER 9

ELEMENTS OF PROJECTIVE GEOMETRY

1. Homogeneous coordinates. A point P of real n-dimensional Euclidean geometry was represented in Chap. 1 by a real n-dimensional vector, and we are accustomed to writing $P = (x_1, \ldots, x_n)$. The numbers $x_1, \ldots, x_n$ are unique if a coordinate system is specified, and we shall call them the non-homogeneous coordinates of P.

It is sometimes convenient to represent P by an $(n + 1)$-dimensional vector

$$\bar{P} = (y_1, \ldots, y_{n+1}),$$

where y_{n+1} is any real nonzero number and $y_i = x_i y_{i+1}$. We then call the vector $\bar{P}$ a set of *homogeneous coordinates* of P. Evidently $\bar{Q} = (z_1, \ldots, z_{n+1})$ is also a set of homogeneous coordinates of the same point P *if and only if* $\bar{Q} = t\bar{P}$ *where* $t \neq 0$. It follows that $P = Q$ if and only if $\bar{P}$ and $\bar{Q}$ are linearly dependent.

We are thus led to the study of the geometry of points P represented by corresponding nonzero $(n + 1)$-dimensional vectors $\bar{P}$ such that two points P and Q coincide if and only if $\bar{P}$ and $\bar{Q}$ are linearly dependent. This is a first postulate of the subject called *projective geometry*. The restriction $y_{n+1} \neq 0$ will be omitted, and we only assume that all $(n + 1)$-dimensional vectors considered are nonzero vectors. The points corresponding to vectors with $y_{n+1} = 0$ are usually interpreted as points at infinity.

In the case $n = 3$, we study all nonzero vectors (x, y, z, t). Then the nonhomogeneous coordinates of the points corresponding to those vectors with $t \neq 0$ are the ratios

$$\frac{x}{t}, \quad \frac{y}{t}, \quad \frac{z}{t}.$$

2. Lines and planes. Let $P_1, \ldots, P_m$ be any m distinct points whose nonhomogeneous coordinates are given by

$$(1) \qquad\qquad P_j = (x_{1j}, \ldots, x_{nj}) \qquad (j = 1, \ldots, m).$$

Then the vectors

(2) $$\bar{P}_j = (x_{1j}, \ldots, x_{nj}, 1) \qquad (j = 1, \ldots, n)$$

are m corresponding sets of homogeneous coordinates. Define $P = (x_1, \ldots, x_n)$ by a set of parametric equations

(3) $$P = P_1 + \xi_2(P_2 - P_1) + \cdots + \xi_m(P_m - P_1)$$

for $m - 1$ independent real parameters $\xi_2, \ldots, \xi_m$. The case $m = 2$ is the case of lines, and the case $m = 3$ is the case of planes. Then if $\bar{P} = (x_1, \ldots, x_n, 1)$ and $\lambda_1, \ldots, \lambda_m$ are defined by

(4) $$\lambda_1 = 1 - (\xi_1 + \cdots + \xi_m), \qquad \lambda_j = \xi_j \quad (j = 1, \ldots, m),$$

we see that

(5) $$\bar{P} = \lambda_1 \bar{P}_1 + \cdots + \lambda_m \bar{P}_m.$$

For $P_j - P_1 = (x_{1j} - x_{11}, \ldots, x_{nj} - x_{n1}, 0)$ and formula (5) is equivalent to

(6) $$\bar{P} - \bar{P}_1 = \xi_2(\bar{P}_2 - \bar{P}_1) + \cdots + \xi_m(\bar{P}_m - \bar{P}_1).$$

The $(n + 1)$st coordinates of the two sides of this vector equation are both zero, and the statement of the equality of the first n coordinates is precisely formula (3).

Conversely, let $\bar{P}$ be defined by formula (5) where the parameters $\lambda_1, \ldots, \lambda_t$ range independently over all real numbers. If

$$t = \lambda_1 + \cdots + \lambda_n$$

is zero, $\bar{P}$ represents a point at infinity. Define

(7) $$j = \frac{\lambda_j}{t} \qquad (j = 2, \ldots, m)$$

for all finite points, and see that

(8) $$\bar{Q} = (x_1, \ldots, x_m, 1) = \frac{1}{t}\bar{P} = \frac{\lambda_1}{t}\bar{P}_1 + \xi_2\bar{P}_2 + \cdots + \xi_m\bar{P}_m.$$

Then $t^{-1}(\lambda_1 - t) = t^{-1}(\lambda_2 + \cdots + \lambda_m) = -(\xi_2 + \cdots + \xi_m)$ and

(9) $$\bar{Q} - \bar{P}_1 = \frac{\lambda_1 - t}{t}\bar{P}_1 + \xi_2\bar{P}_2 + \cdots + \xi_m\bar{P}_m$$
$$= \xi_2(\bar{P}_2 - \bar{P}_1) + \cdots + \xi_m(\bar{P}_m - \bar{P}_1).$$

Hence, formula (3) holds with $P = (x_1, \ldots, x_n)$. Note that if $\bar{R}_i = t_i\bar{P}_i$, then formula (5) is equivalent to $\bar{P} = \mu_1 R_1 + \cdots + \mu_m \bar{R}_m$, where $\mu_i \bar{R}_i = \lambda_i \bar{P}_i = \mu_i t_i \bar{P}_i$. Then $\lambda_i = \mu_i t_i$ with $t_i \neq 0$, and the parameters μ_i range over all real numbers when the parameters λ_i do. Hence, formula (3) is equivalent to formula (5) where $\bar{P}_i$ is any vector of homogeneous coordinates of P_i.

We have now shown that the geometric configuration defined by the parametric equations of formula (3) is also defined by the equations of formula (5). However, formula (5) is homogeneous in its parameters and is a much more convenient form for the parametric equations of what are called *linear spaces* in geometry.

In case $m = 2$ and P_1 and P_2 are distinct points, the corresponding vectors $\bar{P}_1$ and $\bar{P}_2$ are linearly independent. Then the line through P_1 and P_2 is the set of points (including points at infinity) whose homogeneous coordinates are all linear combinations of P_1 and P_2. In the case $n = 3$, the homogeneous coordinate equations of a line become

$$(10) \qquad (x, y, z, t) = \alpha(x_1, y_1, z_1, t_1) + \beta(x_2, y_2, z_2, t_2)$$

for independent parameters α and β.

Three noncollinear points P_1, P_2, P_3 determine a plane. Then $\bar{P}_1$ and $\bar{P}_2$ must be linearly independent and $\bar{P}_3$ must not be a linear combination of $\bar{P}_1$ and $\bar{P}_2$. It follows that P_1, P_2, P_3 determine a plane if and only if $\bar{P}_1, \bar{P}_2, \bar{P}_3$ are a set of three linearly independent vectors. Then the equations

$$\bar{P} = \alpha\bar{P}_1 + \beta\bar{P}_2 + \gamma\bar{P}_3$$

are a set of parametric equations of the plane. In the case $n = 3$, these equations become

$$(x, y, z, t) = \alpha(x_1, y_1, z_1, t_1) + \beta(x_2, y_2, z_2, t_2) + \gamma(x_3, y_3, z_3, t_3).$$

We shall limit all further study to the case $n = 3$.

EXERCISES

1. Give a set of parametric equations in homogeneous coordinates of the lines defined by following pairs of points:

(a) $(1, -1, 1), (3, -3, 2)$

(b) $(1, 2, 3), (0, 2, 3)$

(c) $(-2, 1, 1), (3, -2, 2)$

(d) $(3, -1, 2), (1, -1, -2)$

(e) $(1, 2, -3), (-3, -2, 1)$

(f) $(4, 1, 6), (2, -1, 4)$

(g) $(1, 0, 2), (2, -1, 1)$

(h) $(1, 1, 1), (2, 2, 5)$

2. Give a set of parametric equations in homogeneous coordinates of the planes defined by the following triples of points:

(a) $(1, -1, 1)$, $(3, -3, 2)$, $(2, 1, 0)$

(b) $(1, 2, 3)$, $(0, 2, 3)$, $(-1, 1, 2)$

(c) $(-2, 1, 1)$, $(3, -2, 2)$, $(0, 1, 0)$

(d) $(3, -1, -2)$, $(1, -1, -5)$, $(-2, -6, 3)$

3. Projective transformations. Let C be any nonsingular 4×4 matrix

$$(11) \qquad C = \begin{pmatrix} c_{11} & c_{12} & c_{13} & c_{14} \\ c_{21} & c_{22} & c_{23} & c_{24} \\ c_{31} & c_{32} & c_{33} & c_{34} \\ c_{41} & c_{42} & c_{43} & c_{44} \end{pmatrix},$$

and define

$$(12) \qquad \bar{Q} = (x', y', z', t') = \bar{P}C$$

for every point P with corresponding homogeneous coordinate vector $P = (x, y, z, t)$. Then every point P is mapped on (*i.e.*, determines) a unique point Q, which we shall call the *image* of P under a projective transformation with matrix C. Conversely, if the image Q is given, P is uniquely determined; for $\bar{P} = \bar{Q}C^{-1}$.

It should be noted that a projective transformation may map finite points on points at infinity. For example,

$$(13) \qquad (1, 0, 0, 1) \begin{pmatrix} 0 & 1 & 0 & 0 \\ 0 & 0 & 1 & 0 \\ 0 & 0 & 0 & 1 \\ 1 & 0 & 0 & 0 \end{pmatrix} = (1, 1, 0, 0).$$

It is natural to ask when two matrices B and C define the same projective transformation. This occurs when and only when $\bar{Q} = \bar{P}B$ and $\bar{R} = \bar{P}C$ define the same point Q for every P. This means that

$$(14) \qquad \bar{P}C = t(\bar{P}B)$$

for every P where t is a nonzero real number which may conceivably depend on P. We shall prove the following theorem.

Theorem 1. *Two matrices* B *and* C *define the same projective transformation if and only if* C *is a nonzero scalar multiple of* B.

For if $C = tB$, it is evident that $PC = t(PB)$. Conversely, suppose that PC is a nonzero scalar multiple of $\bar{P}B$ for every P. Take $P = (1, 0, 0, 0)$ and see that $\bar{P}B$ is the first row B_1 of B. Then the first row of C must be a scalar multiple t_1B_1 of the first row of B. We have a similar result for the other rows and see that $C = DB$, where $D = diag\{t_1, t_2, t_3, t_4\}$ is a diagonal matrix. We next take $P = (1, 1, 1, 1)$ and see that

$$(15) \quad \bar{P}C = t_1B_1 + t_2B_2 + t_3B_3 + t_4B_4 = t\bar{P}B$$
$$= t(B_1 + B_2 + B_3 + B_4)$$

where B_i is the ith row of A. Then

$$(16) \qquad (t_1 - t, t_2 - t, t_3 - t, t_4 - t)B = 0.$$

Since B is nonsingular, $t_1 = t_2 = t_3 = t_4 = t$, and we have proved that $C = tB$.

4. Tetrahedral coordinates. If P_1 is any point, the corresponding vector P_1 is a nonzero vector with the property that all nonzero scalar multiples of $\bar{P}_1$ define the same point P_1. Let P_2 be a point distinct from P_1. Then P_1 and P_2 are linearly independent, and the set of all linear combinations of $\bar{P}_1$ and $\bar{P}_2$ is the set of all vectors P defining points on the line joining P_1 to P_2.

We next let P_3 be a point not on the line joining P_1 and P_2. Then $\bar{P}_1, \bar{P}_2, \bar{P}_3$ are linearly independent, and the set of all linear combinations of $\bar{P}_1, \bar{P}_2, \bar{P}_3$ is the set of all vectors $\bar{P}$ defining points P on the plane determined by P_1, P_2, P_3.

Suppose, finally, that P_4 is a point not on the plane determined by P_1, P_2, P_3. Then $\bar{P}_4$ is not a linear combination of $\bar{P}_1, \bar{P}_2, \bar{P}_3$. But then $\bar{P}_1, \bar{P}_2, \bar{P}_3, \bar{P}_4$ are linearly independent. It follows that the matrix

$$(17) \qquad B = \begin{pmatrix} \bar{P}_1 \\ \bar{P}_2 \\ \bar{P}_3 \\ \bar{P}_4 \end{pmatrix},$$

whose rows are the vectors $\bar{P}_1, \bar{P}_2, \bar{P}_3, \bar{P}_4$, is a nonsingular matrix. For $|B| = 0$ if and only if there exist numbers $\lambda_1, \lambda_2, \lambda_3, \lambda_4$ not all zero such that $\lambda_1\bar{P}_1 + \lambda_2\bar{P}_2 + \lambda_3\bar{P}_3 + \lambda_4\bar{P}_4 = 0$.

Let P_1, P_2, P_3, P_4 be four fixed points which are not coplanar and let C be the nonsingular matrix formed by the corresponding vectors $\bar{P}_1, \bar{P}_2, \bar{P}_3, \bar{P}_4$. Then every vector $\bar{P} = (x, y, z, t)$ defines a unique set of numbers (x', y', z', t') such that

(18) $$\bar{P} = x'\bar{P}_1 + y'\bar{P}_2 + z'\bar{P}_3 + t'\bar{P}_4.$$

We shall call these numbers a set of *tetrahedral coordinates* of P and shall say that P_1, P_2, P_3, P_4 are the vertices of a tetrahedron of reference. The existence of such coordinates is due to the fact that formula (18) is equivalent to

(19) $$(x, y, z, t) = (x', y', z', t')B,$$

and therefore to

(20) $$(x', y', z', t') = (x, y, z, t)C,$$

where $C = B^{-1}$.

We have been using the points whose homogeneous coordinates are $(1, 0, 0, 1)$, $(0, 1, 0, 1)$, $(0, 0, 1, 1)$ and $(0, 0, 0, 1)$ as a tetrahedron of reference and we see that $(x, y, z, t) = x(1, 0, 0, 1) + y(0, 1, 0, 1) + z(0, 0, 1, 1) + (t - x - y - z)(0, 0, 0, 1)$. However, x, y, z, t are actually the tetrahedral coordinates of a point relative to the tetrahedron defined by the vectors $(1, 0, 0, 0)$, $(0, 1, 0, 0)$, $(0, 0, 1, 0)$, $(0, 0, 0, 1)$.

The equations of formula (20) have been interpreted in this section as the relations connecting two sets of coordinates of a fixed point. They were given in formula (12) of Sec. 3 as defining a point-to-point correspondence called a *projective transformation*. It is important to observe that these two interpretations are two geometric interpretations of the same algebraic phenomenon.

We should also note that if L is a three-rowed orthogonal matrix and

(21) $$C = \begin{pmatrix} L & 0 \\ 0 & 1 \end{pmatrix},$$

then the equations

(22) $$(x, y, z, 1) = (x', y', z', 1)\begin{pmatrix} L & 0 \\ 0 & 1 \end{pmatrix}$$

define a projective transformation of coordinates with orthogonal matrix C. But these equations are clearly equivalent to the orthogonal transformation

(23) $$(x, y, z) = (x', y', z')L$$

on nonhomogeneous coordinates.

5. The unit point. As we have seen, any set of four points P_1, P_2, P_3, P_4 that are not coplanar defines a corresponding matrix B of formula (17) and hence a set of tetrahedral coordinates. However, the same points also define a matrix

(24)
$$\begin{pmatrix} t_1 & 0 & 0 & 0 \\ 0 & t_2 & 0 & 0 \\ 0 & 0 & t_3 & 0 \\ 0 & 0 & 0 & t_4 \end{pmatrix} B = \begin{pmatrix} t_1 P_1 \\ t_2 P_2 \\ t_3 P_3 \\ t_4 P_4 \end{pmatrix}$$

for any set of nonzero real numbers $t_1, \ldots, t_4$. It follows that $P_1, \ldots, P_4$ do not completely specify B. The specifcation may be completed however by prescribing a fixed point (x_0, y_0, z_0, t_0) as a point called a *unit point* and which is such that $x_0' = y_0' = z_0' = t_0' = 1$. For then we have

(25)
$$(x, y, z, t) = t(1, 1, 1, 1) \begin{pmatrix} t_1 P_1 \\ t_2 P_2 \\ t_3 P_3 \\ t_4 P_4 \end{pmatrix}$$

which is equivalent to

(26) $$t(t_1, t_2, t_3, t_4)B = (x_0, y_0, z_0, t_0)$$

and therefore to

(27) $$t(t_1, t_2, t_3, t_4) = (x_0, y_0, z_0, t_0)C.$$

But this uniquely determines the numbers (t_1, t_2, t_3, t_4) apart from a proportionality factor t.

Note that the unit point is not arbitrary since we must have t_1, t_2, t_3, t_4 all not zero. But the equation $t(t_1, t_2, t_3, t_4)B = (x_0, y_0, z_0, t_0)$, with one of t_1, t_2, t_3, t_4 zero, implies that (x_0, y_0, z_0, t_0) is a linear combination of three of the vectors $\bar{P}_1, \bar{P}_2, \bar{P}_3, \bar{P}_4$ and therefore P_0 is a point on a plane through three of the vertices of the tetrahedron of reference. The four planes determined by selecting three of the four vertices are called the *faces* of the tetrahedron of reference, and we have shown that *a point may be selected as unit point if and only if it is not a point of a face of the tetrahedron.*

EXERCISES

1. Let the vertices of a tetrahedron of reference be $P_1 = (\frac{2}{3}, -\frac{1}{3}, \frac{2}{3})$, $P_2 = (\frac{2}{3}, \frac{2}{3}, -\frac{1}{3})$, $P_3 = (-\frac{1}{3}, \frac{2}{3}, \frac{2}{3})$, $P_4 = (-\frac{1}{6}, \frac{1}{3}, \frac{1}{3})$ and the unit point

be $(2, 1, 1)$. Find the tetrahedral coordinates of the points whose homogeneous coordinates are given by the following vectors:

(a) $(1, 0, 0, 1)$		(e) $(1, -1, 2, 0)$
(b) $(1, 1, 1, 1)$		(f) $(4, 1, -1, 1)$
(c) $(1, 0, 0, 0)$		(g) $(-1, 2, -1, 1)$
(d) $(0, 0, 0, 1)$		(h) $(1, 2, 2, 1)$

2. Let the vectors of Exercise 1 be the tetrahedral coordinates of a set of points. Find the (nonhomogeneous) rectangular coordinates of the finite points of the set.

6. Invariant points. A projective transformation with matrix C maps every point P on a unique image Q where $\bar{Q} = \bar{P}C$. Then we call P an *invariant point* if $Q = P$, that is, P is its own image. But this occurs if and only if $\bar{P}C = t\bar{P}$, that is,

$$(28) \qquad\qquad \bar{P}(tI - C) = 0.$$

Since $\bar{P}$ must be a nonzero vector, the determinant

$$(29) \qquad\qquad |tI - C| = 0.$$

Thus t must be a *characteristic root* of the matrix C, and we have shown that P *is an invariant point if and only if* $\bar{P}$ *is a characteristic vector of* C.

If t is a simple root of $xI - C = 0$, an argument like that used in the proof of Theorem 17 of Chap. 6 may be used to prove that there is only a single invariant point. When t is a double root of $xI - C = 0$, there are two corresponding linearly independent vectors and all linear combinations of them are characteristic vectors. Then there is a *line of invariant points* corresponding to the root t. The root t may be a triple root of $xI - C = 0$ and then there are three linearly independent characteristic vectors and a corresponding *plane of invariant points*. In the final case of a root of multiplicity four, $C = tI$ and *all points* are invariant points.

7. Quadric surfaces. An algebraic surface of degree n is defined by a polynomial equation $f(x, y, z) = 0$ where $f(x, y, z)$ is a polynomial of degree n. Then the equation

$$(30) \qquad\qquad F(x, y, z, t) \equiv t^n f\left(\frac{x}{t}, \frac{y}{t}, \frac{z}{t}\right) = 0$$

is an equation in homogeneous coordinates of the same surface.

Moreover, the polynomial $F(x, y, z, t)$ is a homogeneous polynomial in x, y, z, t and this is the source of the "homogeneous" coordinate terminology.

A projective transformation $(x, y, z, t) = (x', y', z', t')B$ replaces $F(x, y, z, t)$ by another homogeneous polynomial and the corresponding equation $\phi(x', y', z', t') = 0$ is an equation in tetrahedral coordinates of the given surface. In particular a quadric surface is defined by a polynomial equation $F(x, y, z, t) = 0$ where

$$(31) \quad F(x, y, z, t) \equiv (x, y, z, t) \begin{pmatrix} a_{11} & a_{12} & a_{13} & a_{14} \\ a_{12} & a_{22} & a_{23} & a_{24} \\ a_{13} & a_{23} & a_{33} & a_{34} \\ a_{14} & a_{24} & a_{34} & a_{44} \end{pmatrix} \begin{pmatrix} x \\ y \\ z \\ t \end{pmatrix}.$$

Thus $F(x, y, z, t) = \bar{P}A\bar{P}^*$, where A is a symmetric matrix. Then

$$(32) \qquad \phi(x', y', z', t') \equiv \bar{Q}BAB^*\bar{Q}^*$$

and the matrix of this quadratic form is the matrix BAB^*.

It can be shown that the matrix B may be selected so that

$$(33) \qquad A_0 = BAB^* = \begin{pmatrix} d_1 & 0 & 0 & 0 \\ 0 & d_2 & 0 & 0 \\ 0 & 0 & d_3 & 0 \\ 0 & 0 & 0 & d_4 \end{pmatrix}$$

where $d_i = 1, -1$, or zero. The corresponding quadratic form is then $d_1x'^2 + d_2y'^2 + d_3z'^2 + d_4t'^2$. We shall not prove this result here but refer the reader to the author's "Introduction to Algebraic Theories" for proof. The number of nonzero diagonal terms is the rank of the quadratic form $F(x, y, z, t)$ and the number of positive diagonal elements d_i is its index. Both integers are invariants of the form. Note that $|A_0| = |B|^2|A|$.

The determinant of the matrix A of the quadratic form $F(x, y, z, t)$ is called a *discriminant* Δ of the corresponding quadric surface, and we have seen that a projective transformation of coordinates replaces Δ by $|B|^2\Delta$, where B is the matrix of the transformation. We call the surface nonsingular if $\Delta \neq 0$ and see that then $|B|^2\Delta = d_1d_2d_3d_4 \neq 0$ and each $d_i = 1$ or -1. By permuting variables and multiplying by -1 if necessary, we see that every nonsingular quadric surface is defined with respect

to a properly chosen tetrahedral coordinate system by one and
only one of the equations

$$(34) \qquad \begin{aligned} x^2 + y^2 + z^2 + t^2 &= 0, \\ x^2 + y^2 + z^2 - t^2 &= 0, \\ x^2 + y^2 - z^2 - t^2 &= 0. \end{aligned}$$

8. Cross ratios of points.   If the rows of the matrix

$$U = \begin{pmatrix} a_1 & a_2 & \cdots & a_n \\ b_1 & b_2 & \cdots & b_n \end{pmatrix}$$

are not proportional, then it must be possible to select two columns of U such that the rows of

$$(35) \qquad \begin{pmatrix} a_i & a_j \\ b_i & b_j \end{pmatrix}$$

are not proportional.   Then these rows are linearly independent
and the only solution of

$$(x_1, x_2) \begin{pmatrix} a_i & a_j \\ b_i & b_j \end{pmatrix} = x_1(a_i, a_j) + x_2(b_i, b_j) = 0$$

is $x_1 = x_2 = 0$.   It follows that the determinant

$$(36) \qquad \Delta = \begin{vmatrix} a_i & a_j \\ b_i & b_j \end{vmatrix} \neq 0.$$

Let us now consider a matrix

$$(37) \qquad A = \begin{pmatrix} x_1 & y_1 & z_1 & t_1 \\ x_2 & y_2 & z_2 & t_2 \\ x_3 & y_3 & z_3 & t_3 \\ x_4 & y_4 & z_4 & t_4 \end{pmatrix}$$

whose rows are the homogeneous coordinates of four distinct
collinear points.   Then we have the following result:

Lemma 1.   *The two-rowed determinants defined by two columns
of* A *are all not zero or all zero.*

For let $U_1 = (a_1, b_1, c_1, d_1)$ and $U_2 = (a_2, b_2, c_2, d_2)$ be the
coordinates of any two distinct points on the line joining the four
given points and so have

$$(38) \qquad (x_i, y_i, z_i, t_i) = \lambda_i U_1 + \mu_i U_2.$$

Then

$$(39) \qquad A = \begin{pmatrix} \lambda_1 & \mu_1 \\ \lambda_2 & \mu_2 \\ \lambda_3 & \mu_3 \\ \lambda_4 & \mu_4 \end{pmatrix} \begin{pmatrix} a_1 & a_2 & a_3 & a_4 \\ b_1 & b_2 & b_3 & b_4 \end{pmatrix}$$

and the two-rowed matrices in the jth and kth columns of A are the two-rowed square submatrices of

$$(40) \qquad \begin{pmatrix} \lambda_1 & \mu_1 \\ \lambda_2 & \mu_2 \\ \lambda_3 & \mu_3 \\ \lambda_4 & \mu_4 \end{pmatrix} \begin{pmatrix} a_j & a_k \\ b_j & b_k \end{pmatrix}.$$

If we let D_{pq} be the determinant formed by the pth and qth rows of this matrix, we have

$$(41) \qquad D_{pq} = \begin{vmatrix} \lambda_p & \mu_p \\ \lambda_q & \mu_q \end{vmatrix} \cdot \begin{vmatrix} a_j & a_k \\ b_j & b_k \end{vmatrix}.$$

If (λ_p, μ_p) and (λ_q, μ_q) are proportional, then so are (x_p, y_p, z_p, t_p) and (x_q, y_q, z_q, t_q) contrary to the hypothesis that the four given points are distinct. It follows that the determinants

$$(42) \qquad \begin{vmatrix} \lambda_p & \mu_p \\ \lambda_q & \mu_q \end{vmatrix}$$

are all not zero and therefore $D_{pq} = 0$ if and only if the determinant

$$(43) \qquad \begin{vmatrix} a_j & a_k \\ b_j & b_k \end{vmatrix},$$

which is independent of p and q, is zero.   This proves the lemma.

The *cross ratio* of four distinct collinear points P_1, P_2, P_3, P_4 is the ratio

$$(44) \qquad k(P_1, P_2, P_3, P_4) = \left(\frac{D_{13}}{D_{14}} \right) \left(\frac{D_{24}}{D_{23}} \right)$$

of the corresponding two-rowed minors D_{pq} obtained from a corresponding matrix

$$A = \begin{pmatrix} \bar{P}_1 \\ \bar{P}_2 \\ \bar{P}_3 \\ \bar{P}_4 \end{pmatrix}$$

of homogeneous coordinates by selecting two columns of A in which the minors D_{pq} are not zero. Then if U_1 and U_2 are the vectors of homogeneous coordinates of any two points on the given line through $P_1, P_2, P_3, P_4,$ we see that

$$(45) \qquad k(P_1, P_2, P_3, P_4) = \frac{\begin{vmatrix} \lambda_1 & \mu_1 \\ \lambda_3 & \mu_3 \end{vmatrix}}{\begin{vmatrix} \lambda_1 & \mu_1 \\ \lambda_4 & \mu_4 \end{vmatrix}} \cdot \frac{\begin{vmatrix} \lambda_2 & \mu_2 \\ \lambda_4 & \mu_4 \end{vmatrix}}{\begin{vmatrix} \lambda_2 & \mu_2 \\ \lambda_3 & \mu_3 \end{vmatrix}}.$$

Then $k(P_1, P_2, P_3, P_4)$ *is independent of the particular columns of* A *selected in its computation.* It is independent of the base vectors U_1 and U_2, since it is not defined in terms of these vectors but rather in terms of homogeneous coordinates of the given points. Finally, it should be noted that the replacement of $\bar{P}_1, \bar{P}_2, \bar{P}_3, \bar{P}_4$ by $t_1\bar{P}_1, t_2\bar{P}_2, t_3\bar{P}_3, t_4\bar{P}_4$ will not alter $k(P_1, P_2, P_3, P_4)$.

Theorem 2. *The cross ratio of four points is unaltered by a projective transformation.*

For let $\bar{Q}_1 = \bar{P}_1C, \bar{Q}_2 = \bar{P}_2C, \bar{Q}_3 = \bar{P}_3C, \bar{Q}_4 = \bar{P}_4C.$ Then we have

$$(46) \qquad \begin{pmatrix} \bar{Q}_1 \\ \bar{Q}_2 \\ \bar{Q}_3 \\ \bar{Q}_4 \end{pmatrix} = AC = \begin{pmatrix} \lambda_1\mu_1 \\ \lambda_2\mu_2 \\ \lambda_3\mu_3 \\ \lambda_4\mu_4 \end{pmatrix} \begin{pmatrix} U_1 \\ U_2 \end{pmatrix} C = \begin{pmatrix} \lambda_1\mu_1 \\ \lambda_2\mu_2 \\ \lambda_3\mu_3 \\ \lambda_4\mu_4 \end{pmatrix} \begin{pmatrix} U_1C \\ U_2C \end{pmatrix}.$$

Since $k(P_1, P_2, P_3, P_4)$ is independent of the selection of base points, the cross ratio obtained with the use of the vectors U_1C, U_2C will have the same value as that obtained by the use of U_1, U_2. Then $k(Q_1, Q_2, Q_3, Q_4) = k(P_1, P_2, P_3, P_4)$ as desired.

The converse of Theorem 2 is true, and we shall prove it. We state the result as the following

Theorem 3. *Let* P_1, P_2, P_3, P_4 *and* Q_1, Q_2, Q_3, Q_4 *be two sets each of four distinct collinear points, and let*

$$k = k(P_1, P_2, P_3, P_4) = k(Q_1, Q_2, Q_3, Q_4).$$

Then there exists a projective transformation carrying Q_i *into* P_i, *for* i = 1, 2, 3, 4.

To prove this result, we first notice that P_1 and P_2 may be used as base points and therefore $\bar{P}_3 = t_1\bar{P}_1 + t_2\bar{P}_2$, where necessarily $t_1t_2 \neq 0$. Then the vector $t_i\bar{P}_i$ is a homogeneous coordinate representation of P_i and may be taken to be $\bar{P}_i$, for $i = 1, 2$.

Hence, we may take $\bar{P}_3 = \bar{P}_1 + \bar{P}_2$. Now $\bar{P}_4 = s_1\bar{P}_1 + s_2\bar{P}_2$, and we see as before that $s_1 s_2 \neq 0$. Then we may replace P_4 by $s_2^{-1}\bar{P}_4$ and thus obtain $\bar{P}_4 = k\bar{P}_1 + P_2$, where k is a nonzero real number. This yields the relation

$$(47) \qquad A = \begin{pmatrix} 1 & 0 \\ 0 & 1 \\ 1 & 1 \\ k & 1 \end{pmatrix} \begin{pmatrix} \bar{P}_1 \\ \bar{P}_2 \end{pmatrix},$$

and we use formula (45) to see that

$$(48) \qquad k(P_1, P_2, P_3, P_4) = \frac{\begin{vmatrix} 1 & 0 \\ 1 & 1 \end{vmatrix} \begin{vmatrix} 0 & 1 \\ k & 1 \end{vmatrix}}{\begin{vmatrix} 1 & 0 \\ k & 1 \end{vmatrix} \begin{vmatrix} 0 & 1 \\ 1 & 1 \end{vmatrix}} = k.$$

By the argument of Sec. 4, there exists a nonsingular matrix G whose first two rows are $\bar{P}_1$ and $\bar{P}_2$ and then

$$(49) \qquad A = \begin{pmatrix} 1 & 0 & 0 & 0 \\ 0 & 1 & 0 & 0 \\ 1 & 1 & 0 & 0 \\ k & 1 & 0 & 0 \end{pmatrix} G.$$

In a similar fashion the vectors $\bar{Q}_1, \bar{Q}_2, \bar{Q}_3, \bar{Q}_4$ may be chosen so that the corresponding matrix

$$B = \begin{pmatrix} \bar{Q}_1 \\ \bar{Q}_2 \\ \bar{Q}_3 \\ \bar{Q}_4 \end{pmatrix} = \begin{pmatrix} 1 & 0 & 0 & 0 \\ 0 & 1 & 0 & 0 \\ 1 & 1 & 0 & 0 \\ k & 1 & 0 & 0 \end{pmatrix} H,$$

where H is a nonsingular matrix whose first two rows are $\bar{Q}_1$ and $\bar{Q}_2$. The value of k is the same and $B = AC$, where $C = G^{-1}H$ is a matrix defining the desired projective transformation. This proves the theorem.

The form of the matrix A given by formula (47) may be used to determine the effect of permuting the points P_1, P_2, P_3, P_4. There are $4! = 24$ possible permutations and thus possibly 24 distinct values. However, there are actually only six formally distinct values. These are the values

$$(50) \quad k, \quad \frac{1}{k}, \quad 1-k, \quad \frac{1}{1-k}, \quad 1-\frac{1}{k} = \frac{k-1}{k}, \quad \frac{k}{k-1}$$

obtained from k by the two operations of inverting k and subtracting k from 1. That these are values follows from the fact that

$$(51) \qquad k(P_2, P_1, P_3, P_4) = \frac{\begin{vmatrix} 0 & 1 \\ 1 & 1 \end{vmatrix} \cdot \begin{vmatrix} 1 & 0 \\ k & 1 \end{vmatrix}}{\begin{vmatrix} 0 & 1 \\ k & 1 \end{vmatrix} \cdot \begin{vmatrix} 1 & 0 \\ 1 & 1 \end{vmatrix}} = \frac{1}{k},$$

and

$$(52) \qquad k(P_1, P_3, P_2, P_4) = \frac{\begin{vmatrix} 1 & 0 \\ 0 & 1 \end{vmatrix} \cdot \begin{vmatrix} 1 & 1 \\ k & 1 \end{vmatrix}}{\begin{vmatrix} 1 & 0 \\ k & 1 \end{vmatrix} \cdot \begin{vmatrix} 1 & 1 \\ 0 & 1 \end{vmatrix}} = 1 - k,$$

so that

$$(53) \qquad \begin{aligned} k(P_3, P_1, P_2, P_4) &= \frac{1}{1 - k}, \\ k(P_2, P_3, P_1, P_4) &= 1 - \frac{1}{k}, \\ k(P_3, P_2, P_1, P_4) &= \frac{k}{k - 1}. \end{aligned}$$

But from each of these six values we obtain three other equal values by permutation, since it should be evident from formula (45) that

$$(54) \quad k(P_1, P_2, P_3, P_4) = k(P_2, P_1, P_4, P_3) = k(P_3, P_4, P_1, P_2) \\ = k(P_4, P_3, P_2, P_1).$$

ILLUSTRATIVE EXAMPLE

Compute the cross ratio of $P_1 = (1, 1, 2)$, $P_2 = (-3, 1, 1)$, $P_3 = (-\frac{7}{5}, 1, \frac{7}{5})$, $P_4 = (-\frac{1}{3}, 1, \frac{5}{3})$ and the cross ratio of $Q_1 = (-\frac{3}{5}, \frac{7}{5}, \frac{2}{5})$, $Q_2 = (-1, 1, 0)$, $Q_3 = (0, 2, 1)$, $Q_4 = (-\frac{1}{3}, \frac{5}{3}, \frac{2}{3})$, and show that the first four points can be carried into a proper arrangement of the second set of four points by a projective transformation.

Solution

We form the matrices

$$A = \begin{pmatrix} 1 & 1 & 2 & 1 \\ -3 & 1 & 1 & 1 \\ -7 & 5 & 7 & 5 \\ -1 & 3 & 5 & 3 \end{pmatrix}, \quad B = \begin{pmatrix} -3 & 7 & 2 & 5 \\ -1 & 1 & 0 & 1 \\ 0 & 2 & 1 & 1 \\ -1 & 5 & 2 & 3 \end{pmatrix}$$

Then we use the first and second columns to compute

$$k(P_1, P_2, P_3, P_4) = \frac{\begin{vmatrix} 1 & 1 \\ -7 & 5 \end{vmatrix} \begin{vmatrix} -3 & 1 \\ -1 & 3 \end{vmatrix}}{\begin{vmatrix} 1 & 1 \\ -1 & 3 \end{vmatrix} \begin{vmatrix} -3 & 1 \\ -7 & 5 \end{vmatrix}} = \frac{12(-8)}{4(-8)} = 3$$

and

$$k(Q_1, Q_2, Q_3, Q_4) = \frac{\begin{vmatrix} -3 & 7 \\ 0 & 2 \end{vmatrix} \begin{vmatrix} -1 & 1 \\ -1 & 5 \end{vmatrix}}{\begin{vmatrix} -3 & 7 \\ -1 & 5 \end{vmatrix} \begin{vmatrix} -1 & 1 \\ 0 & 2 \end{vmatrix}} = \frac{(-6)(-4)}{(-8)(-2)} = \frac{3}{2}.$$

Then $k(Q_1, Q_2, Q_3 \, Q_4) = 3$, and there exists a projective transformation that will carry P_1 into Q_3, P_2 into Q_2, P_3 into Q_1 and P_4 into Q_4.

EXERCISES

1. Compute $k(P_1, P_2, P_3, P_4)$ and $k(Q_1, Q_2, Q_3, Q_4)$ in the following cases, and find an arrangement of the second set of points into which the first set can be carried by a projective transformation.

(a) $P_1 = (2, 1, -3)$, $P_2 = (-1, 2, 0)$, $P_3 = (-1, -5, -3)$, $P_4 = (4, -3, -3)$

$Q_1 = (1, -2, -1)$, $Q_2 = (-1, 0, 1)$, $Q_3 = (-1, 1, 1)$, $Q_4 = (-4, 3, 4)$

(b) $P_1 = (1, 1, 0)$, $P_2 = (-1, 1, 1)$, $P_3 = (-1, 3, 2)$, $P_4 = (2, 0, -1)$

$Q_1 = (2, -1, 1)$, $Q_2 = (1, 1, -1)$, $Q_3 = (5, -1, 1)$, $Q_4 = (-1, 2, -2)$

(c) $\bar{P}_1 = (1, -1, 1, 1)$, $\bar{P}_2 = (1, 0, -1, 2)$, $\bar{P}_3 = (0, -1, 2, -1)$, $\bar{P}_4 = (5, -1, -3, 9)$

$\bar{Q}_1 = (2, 1, -1, 0)$, $\bar{Q}_2 = (6, 5, 3, 4)$, $\bar{Q}_3 = (1, 1, 1, 1)$, $\bar{Q}_4 = (1, 0, -2, -1)$

(d) $\bar{P}_1 = (0, 1, 0, 1)$, $\bar{P}_2 = (1, 1, 1, 1)$, $\bar{P}_3 = (2, 5, 2, 5)$, $\bar{P}_4 = (1, 3, 1, 3)$

$\bar{Q}_1 = (1, 1, 0, 0)$, $\bar{Q}_2 = (2, 3, 1, 0)$, $\bar{Q}_3 = (0, 1, 1, 0)$, $\bar{Q}_4 = (7, 9, 2, 0)$

(e) $\bar{P}_1 = (0, 1, 0, 4)$, $\bar{P}_2 = (2, 3, -2, 2)$, $\bar{P}_3 = (1, 1, -1, -1)$, $\bar{P}_4 = (1, 2, -1, 3)$

$\bar{Q}_1 = (0, 1, 0, 4)$, $\bar{Q}_2 = (2, 3, -2, 2)$, $\bar{Q}_3 = (2, 4, -2, 6)$, $\bar{Q}_4 = (-2, -2, 2, 2)$

(f) $\bar{P}_1 = (2, 0, 0. 1)$, $\bar{P}_2 = (1, 2, 3, 4)$, $\bar{P}_3 = (-1, 2, 3, 3)$, $\bar{P}_4 = (5, 2, 3, 6)$

$\bar{Q}_1 = (2, 0, 0, 1)$, $\bar{Q}_2 = (5, 2, 3, 6)$, $\bar{Q}_3 = (-3, 2, 3, 5)$, $\bar{Q}_4 = (9, 2, 3, 8)$

2. Use the method of proof of Theorem 3 to show that a projective transformation can be found which will carry any five points $P_1, \ldots,$

P_5, no four of which are coplanar, into any other five noncoplanar points. HINT: Show that $P_1, \ldots, P_5$ can be carried into $(1, 0, 0, 0)$, $(0, 1, 0, 0)$, $(0, 0, 1, 0)$, $(0, 0, 0, 1)$, $(1, 1, 1, 1)$.

9. Plane coordinates and duality. The coefficients of the homogeneous coordinate equation $ax + by + cz + dt = 0$ form a nonzero vector (a, b, c, d) that may be regarded as a set of coordinates of the corresponding plane. Clearly two nonzero vectors define the same plane if and only if each is a nonzero real scalar multiple of the other, and thus we have a new coordinate system for our projective geometry. We may then express all our projective geometric properties in terms of plane coordinate vectors, rather than point coordinate vectors. The reader should carry this out for the material of Sec. 2.

A point, represented in point coordinates by (x, y, z, t), is said to be *incident* with a plane, represented in plane coordinates by (a, b, c, d), when the point lies on the plane (the plane contains the point). This occurs if and only if the inner product of the two corresponding vectors vanishes, *i.e.*, if and only if

$$(x, y, z, t) \cdot (a, b, c, d) = ax + by + cz + dt = 0.$$

Since the incidence relation is unaltered by the interchange of the point and plane vectors, we have a property of our geometry that we may state as the following principle:

DUALITY PRINCIPLE. *If the words point and plane are interchanged in any true statement about the incidence of points and planes, the result is a true statement.*

We may also formulate this property by saying that if a geometric configuration formed out of points and planes has certain incidences the configuration obtained by the interchange of points and planes will have the same incidences.

The Duality Principle may be extended to linear subspaces of an n-dimensional space. The precise results may be found in more advanced texts on projective geometry.

EXERCISES

1. State the dual of the theorem that three points not on a line determine a plane.

2. Find some other incidence relation, and state its dual.

INDEX

A

Addition, of matrices, 78
  of vectors, 15
Adjoint matrix, 87
Algebraic curve and surface, 40
Angle, between line and plane, 32
  between lines, 17
  off nose, 132
  between planes, 30
  between rays, 7
  between spheres, 58
  between vectors, 4, 13, 125
Axes, coordinate, 8
  of ellipsoid, 59
  rotation of, 107, 136
  translation of, 13
Azimuth, 124

B

Broken line, 6

C

Cartesian coordinates, 7
Center, 59, 64, 66
Characteristic determinant, 95
Characteristic equation, 96
Characteristic roots, 75, 96
Characteristic vectors, 101
Charts, gnomonic, 141
Classification of quadrics, 75, 151
Cofactor, 86
Columns of a matrix, 77
Cone, 40
  quadric, 62
Conic section, 59
Contour graph, 127
Coordinate axes, 8

Coordinate planes, 9
Coordinates, plane, 156
  homogeneous, 143–145
  rectangular, 7
  spherical, 123, 131–133
  stabilization of, 137
  tetrahedral, 147
  of vector, 1
Cosines, direction, 11
Cross ratio, 152–157
Curve, 38
  algebraic, 40
Cylinder, 43
  elliptic, 74
  hyperbolic, 75
  lines on, 48

D

Dependence, linear, 2
Determinants, 87–90
Diagonal matrix, 82
Difference of vectors, 1
Direction angles, 11
Direction cosines, 11
Direction numbers, 16
Distance, between line and point, 23
  between plane and point, 29
  between points, 16
Division of line segment in given
  ratio, 19
Duality Principle, 158

E

Elevation, 124
Ellipsoid, 59
  imaginary, 61
Elliptic cylinder, 74
Elliptic paraboloid, 71

Equation, of a plane, 25
  intercept form, 25
  normal form, 24
  characteristic, 96
Equations of a line, 18
  in symmetric form, 20
Equivalence, orthogonal, 100

F

Families of spheres, 56

G

Gnomonic charts, 141
Gnomonic projections, 139

H

Homogeneous coordinates, 143–145
Homogeneous polynomials, 41
Homogeneous systems of equations,
  94
Hyperbolic cylinder, 75
Hyperbolic paraboloid, 72
Hyperboloid, 63–65
  center of, 64, 66
  lines on, 67

I

Identity matrix, 82
Imaginary ellipsoid, 61
Imaginary sphere, 54
Imaginary surface, 37
Independence, linear, 2
Inner product, 3
Intersection, of lines and surfaces, 47
  of two surfaces, 38
Intercept form of plane equation, 25
Intercepts, 25
Irreducible surface, 40

L

Laws for vectors, 1
Length, of line segment, 16
  signed, 5
  of vector, 3, 11

Linear combination, 2
Linear systems, 92
Line, broken, 6
  equations of, 18, 20
  of intersection, 31
  normal, 51
  and plane, angle between, 32
  and point, angle between, 23
Lines, angle between, 17
  on cylinder, 48
  in homogeneous coordinates, 144
  on hyperboloid, 67
  through origin, 10
  on paraboloid, 73
  and surfaces, intersection of, 47
  of symmetry, 121–122
  tangent, 49

M

Major, minor, mean semiaxes, 59
Matrices of planar rotations, 133–
  136
Matrix, 77
  addition, 78
  adjoint, 87
  determinant of, 87
  identity, 82
  inverse of, 91
  multiplication, 79, 83
  nonsingular, 90
  orthogonal, 98
  partitioning of, 84
  of product of linear transforma-
    tions, 106
  scalar, 82
  scalar product for, 78
  similar to a diagonal matrix, 96
  symmetric, 97

N

n-dimensional vector, 1
Nonsingular matrix, 90
Norm of vector, 3
Normal form of equation of plane, 24
Normal line, 51
Numbers, direction, 16

O

Oblate spheroid, 60
Octants, 9
Ordinary point, 50
Origin, 7
  lines through, 10
Orthogonal equivalence, 100
Orthogonal matrices, 98
Orthogonal projection, 6
Orthogonal reduction of quadratic
  form, 110
Orthogonal transformations, 104
Orthogonal vectors, 4

P

Paraboloids, 71–72
Parallax, 127
Parallel pencil, 33
Parallel planes, 27
Parametric equations, 18, 36
Partitioning of matrices, 84
Pencils of planes, 32
Pitch, 133
Planar rotations, 107
  matrices of, 133–136
Plane, 24–36
  radical, 57
  tangent, 50
  through three points, 27
Plane coordinates, 156
Plane section, 38, 117
Planes, coordinate, 9
  in homogeneous coordinates, 144
  line of intersection of, 31
  parallel, 27
  parametric equations of, 35
  pencils of, 32
  of symmetry, 118–121
Point, and line, distance between, 23
  ordinary, 50
  and plane, distance between, 29
  singular, 50
Point ellipsoid, 61
Point sphere, 54
Point surface, 37

Polynomials, reducible, 40
  homogeneous, 41
Principal axis theorem, 100
Principle of Duality, 158
Products of matrices, 79, 83
  of transformations, 106
Projection, gnomonic, 139
  orthogonal, 7
Projective transformations, 146
Prolate spheroid, 60

Q

Quadratic form, orthogonal reduction of, 110
Quadric cone, 62
Quadric surface, 40
  classifications of, 75, 151
  in projective space, 150
  orthogonal reduction of, 114
  plane sections of, 117
  points of symmetry of, 117

R

Ray, 5
Radical plane, 57
Range, 120
Real vector, 1
Rectangular coordinates, 7
Reducible polynomial and surface,
  40
Reflection of axes, 106
Reguli, 68, 73
Roll, 133
Roots, characteristic, 75, 96
Rotation of axes, 107
  as product of planar rotations, 136

S

Scalar matrix, 82
Scalar product, 2, 79
Sections, of cone, 62
  of ellipsoid, 59
  of hyperboloid, 64, 66
  of paraboloid, 71–72

Semiaxes of an ellipsoid, 59
Signed length, 5
Similar matrices, 96
Singular point, 50
Space curve, 38
Sphere, 54
Spheres, families of, 57
Spherical coordinates, 123, 131–133
Spheroid, oblate, 60
  prolate, 60
Square matrix, 77
Stabilization of coordinates, 137
Sum, of matrices, 78
  of vectors, 1
Surface, 37
  algebraic, 40
  of revolution, 45, 60
  symmetries of, 46
Symmetric equations, 20
Symmetric matrix, 97
Symmetries, 46, 117–122

T

Tangent to quadrics, 52
Tangent lines, 49
Tangent plane, 50
Tetrahedral coordinates, 147
  unit point of, 149
Tetrahedron of reference, 103

Transformations, orthogonal, 104
  projective, 146
Transpose, 81
Translation of axes, 13

U

Unit vector, 3, 101

V

Vectors, 1–4
  addition, 15
  angle between, 4, 13, 125
  inner product, 3
  length, 3, 10
  linear independence of, 2
  norm, 3
  orthogonal, 4
  unit, 3, 101
Vertices of hyperboloid, 66
Vertex of cone, 40
  of paraboloid, 72

Y

Yaw, 133

Z

Zenith, 124
Zero vector, 1